Profiles in Cardiac Pacing and Electrophysiology

Berndt Lüderitz

MD, PhD, MD (Hon.), FESC, FACC, FAHA
Professor and Chairman
Department of Medicine—Cardiology
University of Bonn
Bonn, Germany

Blackwell
Futura

© 2005 Berndt Lüdertiz, MD
Published by Blackwell Publishing
Blackwell Futura is an imprint of Blackwell Publishing

Blackwell Publishing, Inc., 350 Main Street, Malden, Massachusetts 02148-5020, USA
Blackwell Publishing Ltd, 9600 Garsington Road, Oxford OX4 2DQ, UK
Blackwell Science Asia Pty Ltd, 550 Swanston Street, Carlton, Victoria 3053, Australia

First published 2005

ISBN: 1-4051-3116-0

Library of Congress Cataloging-in-Publication Data

Lüderitz, Berndt, 1940–
    Profiles in cardiac pacing and electrophysiology / by Berndt Lüderitz.
  — 1 ed.
        p. ; cm.
    Includes bibliographical references and index.
    ISBN 1-4051-3116-0 (alk. paper)
    1. Arrhythmia—History.   2. Cardiologists—Encyclopedias.   I. Title.
    [DNLM: 1. Arrhythmia—Biography.   2. Arrhythmia—history.
  3. Physicians—Biography.   4. Cardiac Pacing, Artificial—history.
  5. Electrophysiology—history.   6. Heart—physiology.   WZ 112.5.C2 L944p 2005]
    RC685.A65L834 2005
    616.1'28'003—dc22

                                                                      2004016930

A catalogue record for this title is available from the British Library

Acquisitions: Steven Korn
Production: Claire Bonnett
Set in 9.5/12pt Palatino by Graphicraft Limited, Hong Kong
Printed and bound in Denmark by Narayana Press, Odder

For further information on Blackwell Publishing, visit our website:
www.blackwellcardiology.com

Notice: The indications and dosages of all drugs in this book have been recommended
in the medical literature and conform to the practices of the general community. The
medications described do not necessarily have specific approval by the Food and Drug
Administration for use in the diseases and dosages for which they are recommended.
The package insert for each drug should be consulted for use and dosage as approved by
the FDA. Because standards for usage change, it is advisable to keep abreast of revised
recommendations, particularly those concerning new drugs.

**Profiles in
Cardiac Pacing and
Electrophysiology**

Dedication

To my wife, Hedwig, and
our children, Florian, Martin, and Stephan

1000051778

Contents

ΟΡΚΟΣ

ΟΜΝΥΜΙ ΑΠΟΛΛΩΝΑ ΙΗΤΡΟΝ ΚΑΙ ΑΣΚΛΗ-
ΠΙΟΝ ΚΑΙ ΥΓΕΙΑΝ ΚΑΙ ΠΑΝΑΚΑΙΑΝ ΚΑΙ
ΘΕΟΥΣ ΠΑΝΤΑΣ ΤΕ ΚΑΙ ΠΑΣΑΣ, ΙΣΤΟΡΑΣ
ΠΟΙΕΥΜΕΝΟΣ ΕΠΙΤΕΛΕΑ ΠΟΙΗΣΕΙΝ ΚΑ-
ΤΑ ΔΥΝΑΜΙΝ ΚΑΙ ΚΡΙΣΙΝ ΕΜΗΝ ΟΡΚΟΝ
ΤΟΝΔΕ ΚΑΙ ΣΥΓΓΡΑΦΗΝ ΤΗΝΔΕ· ΗΓΗΣΑ
ΣΘΑΙ ΜΕΝ ΤΟΝ ΔΙΔΑΞΑΝΤΑ ΜΕ ΤΗΝ ΤΕΧ
ΝΗΝ ΤΑΥΤΗΝ ΙΣΑ ΓΕΝΕΤΗΣΙΝ ΕΜΟΙΣΙ
ΚΑΙ ΒΙΟΥ ΚΟΙΝΩΣΑΣΘΑΙ ΚΑΙ ΧΡΕΩΝ ΧΡΗΙ-
ΖΟΝΤΙ ΜΕΤΑΔΟΣΙΝ ΠΟΙΗΣΑΣΘΑΙ, ΚΑΙ ΓΕ-
ΝΟΣ ΤΟ ΕΞ ΑΥΤΟΥ ΑΔΕΛΦΟΙΣ ΙΣΟΝ ΕΠΙ-
ΚΡΙΝΕΙΝ ΑΡΡΕΣΙ ΚΑΙ ΔΙΔΑΞΕΙΝ ΤΗΝ ΤΕΧ-
ΝΗΝ ΤΑΥΤΗΝ, ΗΝ ΧΡΗΙΖΩΣΙ ΜΑΝΘΑΝΕΙΝ
ΑΝΕΥ ΜΙΣΘΟΥ ΚΑΙ ΣΥΓΓΡΑΦΗΣ, ΠΑΡΑΓΓΕ-
ΛΙΗΣ ΤΕ ΚΑΙ ΑΚΡΟΗΣΙΟΣ ΚΑΙ ΤΗΣ ΛΟΙΠΗΣ
ΑΠΑΣΗΣ ΜΑΘΗΣΙΟΣ ΜΕΤΑΔΟΣΙΝ ΠΟΙΗΣΑ
ΣΘΑΙ ΥΙΟΙΣΙ ΤΕ ΕΜΟΙΣ ΚΑΙ ΤΟΙΣ ΤΟΥ ΕΜΕ
ΔΙΔΑΞΑΝΤΟΣ ΚΑΙ ΜΑΘΗΤΗΙΣΙ ΣΥΓΓΕΓΡΑ
ΜΕΝΟΙΣΙ ΤΕ ΚΑΙ ΩΡΚΙΣΜΕΝΟΙΣ ΝΟΜΩΙ
ΙΗΤΡΙΚΩΙ, ΑΛΛΩΙ ΔΕ ΟΥΔΕΝΙ. ΔΙΑΙΤΗΜΑ
ΣΙ ΤΕ ΧΡΗΣΟΜΑΙ ΕΠ' ΩΦΕΛΕΙΗΙ ΚΑΜΝΟΝ-
ΤΩΝ ΚΑΤΑ ΔΥΝΑΜΙΝ ΚΑΙ ΚΡΙΣΙΝ ΕΜΗΝ,
ΕΠΙ ΔΗΛΗΣΕΙ ΔΕ ΚΑΙ ΑΔΙΚΙΗΙ ΕΙΡΞΕΙΝ.
ΟΥ ΔΩΣΩ ΔΕ ΟΥΔΕ ΦΑΡΜΑΚΟΝ ΟΥΔΕΝΙ

ΑΙΤΗΘΕΙΣ ΘΑΝΑΣΙΜΟΝ ΟΥΔΕ ΥΦΗΓΗΣΟ
ΜΑΙ ΣΥΜΒΟΥΛΙΗΝ ΤΟΙΗΝΔΕ· ΟΜΟΙΩΣ ΔΕ
ΟΥΔΕ ΓΥΝΑΙΚΙ ΠΕΣΣΟΝ ΦΘΟΡΙΟΝ ΔΩΣΩ.
ΑΓΝΩΣ ΔΕ ΚΑΙ ΟΣΙΩΣ ΔΙΑΤΗΡΗΣΩ ΒΙΟΝ
ΤΟΝ ΕΜΟΝ ΚΑΙ ΤΕΧΝΗΝ ΤΗΝ ΕΜΗΝ. ΟΥ
ΤΕΜΕΩ ΔΕ ΟΥΔΕ ΜΗΝ ΛΙΘΙΩΝΤΑΣ, ΕΚΧΩ
ΡΗΣΩ ΔΕ ΕΡΓΑΤΗΙΣΙΝ ΑΝΔΡΑΣΙ ΠΡΗΞΙ-
ΟΣ ΤΗΣΔΕ. ΕΣ ΟΙΚΙΑΣ ΔΕ ΟΚΟΣΑΣ ΑΝ ΕΣΙΩ
ΕΣΕΛΕΥΣΟΜΑΙ ΕΠ' ΩΦΕΛΕΙΗΙ ΚΑΜΝΟΝ-
ΤΩΝ, ΕΚΤΟΣ ΕΩΝ ΠΑΣΗΣ ΑΔΙΚΙΗΣ ΕΚΟΥ-
ΣΙΗΣ ΚΑΙ ΦΘΟΡΙΗΣ, ΤΗΣ ΤΕ ΑΛΛΗΣ ΚΑΙ
ΑΦΡΟΔΙΣΙΩΝ ΕΡΓΩΝ ΕΠΙ ΤΕ ΓΥΝΑΙΚΕΙ-
ΩΝ ΣΩΜΑΤΩΝ ΚΑΙ ΑΝΔΡΩΙΩΝ, ΕΛΕΥΘΕ-
ΡΩΝ ΤΕ ΚΑΙ ΔΟΥΛΩΝ. Α ΔΕ ΑΝ ΕΝ ΘΕΡΑ
ΠΗΙΗΙ Η ΙΔΩ Η ΑΚΟΥΣΩ Η ΚΑΙ ΑΝΕΥ ΘΕ
ΡΑΠΗΙΗΣ ΚΑΤΑ ΒΙΟΝ ΑΝΘΡΩΠΩΝ, Α ΜΗ
ΧΡΗ ΠΟΤΕ ΕΚΛΑΛΕΙΣΘΑΙ ΕΞΩ, ΣΙΓΗΣΟΜΑΙ,
ΑΡΡΗΤΑ ΗΓΕΥΜΕΝΟΣ ΕΙΝΑΙ ΤΑ ΤΟΙΑΥΤΑ.
ΟΡΚΟΝ ΜΕΝ ΟΥΝ ΜΟΙ ΤΟΝΔΕ ΕΠΙΤΕ-
ΛΕΑ ΠΟΙΕΟΝΤΙ, ΚΑΙ ΜΗ ΣΥΓΧΕΟΝΤΙ, ΕΙΗ
ΕΠΑΥΡΑΣΘΑΙ ΚΑΙ ΒΙΟΥ ΚΑΙ ΤΕΧΝΗΣ ΔΟ-
ΞΑΖΟΜΕΝΩΙ ΠΑΡΑ ΠΑΣΙΝ ΑΝΘΡΩΠΟΙΣ
ΕΣ ΤΟΝ ΑΙΕΙ ΧΡΟΝΟΝ· ΠΑΡΑΒΑΙΝΟΝΤΙ ΔΕ
ΚΑΙ ΕΠΙΟΡΚΟΥΝΤΙ ΤΑΝΑΝΤΙΑ ΤΟΥΤΕΩΝ.

the oath of hippocrates

I swear by Apollo the physician, and Æsculapius, and Hygeia, and Panacea, and all the gods and goddesses, that according to my ability and judgment, I will keep this Oath and its stipulation–to reckon him who taught me this Art equally dear to me as my parents, to share my substance with him, and to relieve his necessities if required; to look upon his offspring in the same footing as my own brothers, and to teach them this Art if they shall wish to learn it, without fee or stipulation; and that by precept, lecture, and every other mode of instruction, I will impart a knowledge of the Art to my own sons, and those of my teachers, and to disciples bound by a stipulation and oath according to the law of medicine, but to none other. I will follow that system of regimen which, according to my ability and judgment, I consider for the benefit of my patients, and abstain from whatever is deleterious and mischievous. I will give no deadly medicine to anyone if asked, nor suggest any such counsel; and in like manner I will not give to a woman a pessary to produce abortion. With purity and with holiness I will pass my life and practice my Art. I will not cut persons laboring under the stone, but will leave this to be done by men who are practitioners of this work. Into whatever houses I enter, I will go into them for the benefit of the sick, and I will abstain from every voluntary act of mischief and corruption; and, further from the seduction of females or males, of freemen and slaves. Whatever, in connection with my professional practice, or not in connection with it, I see or hear, in the life of men, which ought not to be spoken of abroad, I will not divulge, as reckoning that all such should be kept secret. While I continue to keep this Oath unviolated, may it be granted to me to enjoy life and the practice of this Art, respected by all men, in all times. But should I trespass and violate this Oath, may the reverse be my lot.

Foreword

Professor Berndt Lüderitz has made a chivalrous effort to acknowledge many, if not most, contributions to the literature of cardiac electrophysiology. Most prior medical/historical communications have given minimal attention to the area of cardiac arrhythmia, and in this regard, the book serves a unique purpose in addressing the discipline of cardiac electrophysiology. Its alphabetical approach, bibliographical subjects, and name index are very valuable. Specific contributions are mentioned in each person's biography.

Clearly, many aspects of the history of a developing field represent moving targets, especially as related to the dynamic, explosive field of cardiac electrophysiology, and there is a lot happening as one uncovers past and present events. Nonetheless, cardiac electrophysiology today is a fairly well-defined discipline because of the creativity, dedication, and hard work of the folks covered in this text. Digging up the classic contributions from often obscure sources also makes this book a must-read for those who want to get the facts straight. A subtle humor is noted at times and makes interesting reading.

This book will be a source to all students of electrophysiology, including the current generation, and the readers will find a variety of aspects covered. I congratulate Professor Lüderitz, who dedicated a lot of time and effort to bring this together. We should all take advantage of this resource.

Masood Akhtar, MD
Milwaukee, USA

Preface

But there is no possible knowledge,
which arrives not from a pre-existent knowledge.
William Harvey (1578–1657)

This new book, entitled *Profiles in Cardiac Pacing and Electrophysiology*, includes an extensive table of the history of the disorders of cardiac rhythm from the sixteenth to the twentieth century, a series of historical pages developed by the author for every issue of the *Journal of Interventional Cardiac Electrophysiology*, reflecting eminent personalities or events in medicine, and the main chapter on biographical sketches followed by a dictionary/glossary of arrhythmias, electrophysiology, and pacing. The "Encyclopedia rhythmologica" is a collection of short biographies of scientists and physicians who have played (or still play) a significant role in improving diagnosis and therapy of heart rhythm disturbances including electrophysiology and pacing. These names were repeatedly found in the speakers or faculty list of pivotal cardiologic, particularly rhythmological congresses, meetings, symposia, and workshops. Altogether, approximately 250 remarkable individuals are described. By its nature, a "mini-lexicon" can make no claim to completeness. Despite the great effort involved in collecting the personal data, some information could not be obtained. The length of the respective CVs varies according to the available information from the literature (including the Internet) or provided by the named individual itself. Furthermore, it should be taken into consideration that the number of names inevitably is based on subjective criteria. While originally filled with biographies of past pioneers of rhythmology, some of whom lived long ago, the encyclopedia does also include many contemporary rhythmologists. In this respect, the list is not statically predetermined, but rather is subject to biographically conditioned dynamics and must be revised and supplemented in definite intervals. To learn more about the achievements of outstanding individuals in the field, reading of our book *History of the Disorders of Cardiac Rhythms* (3rd edn, Futura, 2002) is suggested. The sources of this collection are, among others, the biographies in the Oral History Mini-Theatre, which regularly takes place during the North

American Society of Pacing and Electrophysiology's Annual Scientific Sessions, and the tributes to the Award Winners who are honored at the annual congresses of the society, now known as the Heart Rhythm Society. Another source is the national electrocardiographical posters that the author compiled for the fiftieth anniversary of the European Society of Cardiology in Amsterdam (August, 2000); and finally the author's private archives. In view of the unavoidable limitation of the list of names, despite all efforts, readers are cordially invited to communicate to the author essential corrections and supplementary requests, including the corresponding biographical data.

Blackwell Futura Publishing Company has provided enthusiastic support in the production of this work. At this time I wish to express my sincere gratitude to Blackwell Futura, especially Gina Almond and Claire Bonnett, as well as to my assistant Hildegard Schilling, without whom the book could not have been published on time.

Berndt Lüderitz (2004)

About the Author

Berndt Lüderitz, MD, PhD, MD (Hon.), FESC, FACC, FAHA is currently Professor of Medicine and Head of the Department of Medicine and Cardiology at the University of Bonn, Bonn, Germany. He is a leading member of many scientific associations, societies, and editorial scientific boards. He has been widely recognized for his work in the field of electrophysiology and history of the disorders of cardiac rhythm. In November 2001, he received the Honorary Doctor degree of the University of Athens, Greece.

Part 1

Encyclopedia
Rhythmologica

Robert Adams

Adams, Robert. (*1791; †1875). Although Robert Adams was primarily a surgeon, he is remembered today mainly for his important contributions to the development of cardiology. He was the son of a Dublin solicitor, and studied at Trinity College and the College of Surgeons. Adams made the connection between a slow pulse and the loss of consciousness, and was the first to realize that cerebral symptoms may be caused by cardiac rhythm disorders. Nearly 20 years later, William Stokes published further observations on this condition, which included a careful analysis of Adams' case. The names of both men are now permanently linked because the condition they discovered is known as "Stokes–Adams syndrome."

Akhtar, Masood. Dr. Akhtar war born in Peshawar, Pakistan, in 1943. He received his undergraduate medical training at the King Edward Medical College in Lahore, Pakistan, and subsequently completed 2 years of clinical training in Lahore. In 1968, he immigrated to the USA, where he carried out landmark research in clinical intracardiac electrophysiology with a special focus on the physiology of the His–Purkinje system. This work led to the first description of the phenomenon of macroreentry within the His–Purkinje system in response to programmed premature ventricular stimulation in patients. In 1977, Akhtar moved to the Sinai–Samaritan Medical Center at the University of Wisconsin Medical School in Milwaukee to serve as Director of Electrocardiography and Clinical Electrophysiology. His research spans a wide range of studies on normal and abnormal clinical cardiac electro-

* = Born
† = Died

Masood Akhtar

physiology, the evaluation and management of unexplained syncope, the demonstration of bundle-branch reentry as a mechanism of sustained monomorphic ventricular tachycardia, the use of radiofrequency catheter ablation of the right bundle branch as a curative intervention in patients with bundle branch reentry tachycardia, modification of a slow atrioventricular (AV) nodal pathway using radiofrequency catheter techniques for elimination of AV node reentry tachycardia, as well as numerous other topics.

Etienne Aliot

Aliot, Etienne. Dr. Aliot graduated from the University of Nancy, France. He became an Associate Professor and then, in 1984, a full Professor of Cardiology. Since 1989 he has been the Chief of the Department of Cardiology at the Hôpital Central, University of Nancy. From 1983 through 1984 he was a Visiting Professor, then an Honorary Professor of Medicine at the University of Oklahoma in Oklahoma City. Since 1994 he has been Chairman of the Nucleus of the Working Group of Arrhythmias of the European Society of Cardiology, for which he is the President of the Working Group on Arrhythmias of the French Society of Cardiology. He is also a fellow of the European Society of Cardiology and a fellow of the American College of Cardiology. His main interests are sudden cardiac death (risk factors, therapy) and interventional electrophysiology (catheter ablation and defibrillation).

Allessie, Maurits A. Beginning with his PhD thesis "Circulating excitation in the heart" in 1977, Dr. Allessie has spent a lifetime tracking electrical excitation in the heart. He is renowned for the painstakingly accurate mapping techniques he uses to follow the wavefront. These procedures, which he conducts *in vitro*, *in vivo*, and on live animals, have helped unlock the secrets of reentrant excitation, led to proof of Moe's hypothesis on circulating wavelets in atrial fibrillation, established a unique form of reentry called "leading circle," and, recently, spearheaded a revolution in atrial fibrillation studies. Allessie and his associates used a goat to explain the many mysteries of atrial fibrillation- and tachycardia-induced atrial remodeling.

Maurits A. Allessie

Anderson, Jeffrey L. Dr. Anderson is a Professor of Internal Medicine (Cardiology) at the University of Utah Medical School and Chief of the Division of Cardiology at LDS Hospital in Salt Lake City, Utah. He is also an Honorary Professor of Medicine at Xi'an Medical University in Xi'an, China. Anderson is active in a number of professional organizations and has held several positions within the American Heart Association, the American College of Cardiology, and the American College of Physicians. He is a prolific writer, and has most recently been published in *Circulation, Journal of the American College of Cardiology, The American Journal of Cardiology,* and *the New England Journal of Medicine.*

Andresen, Dietrich. Dr. Andresen was born in St. Annen, Germany, on March 25, 1948. He was licensed as a specialist in Internal Medicine at the Free University of Berlin (1982), and became Senior Cardiologist at the Interdisciplinary Intensive Care Unit at the same institution, a title held from 1982 to 1983. In 1984, he received the title "Cardiologist" and also became Assistant Medical Director. In 1997, he was appointed Head of the Department of Cardiology and Intensive Care Unit at Vivantes Klinikum Am Urban, Berlin. Then, in October 2000, he became Head of the Department of Cardiology, Vivantes Kliniken Friedrichshain, Berlin. Andresen's main fields of scientific interest are the development and assessment of long-term electrocardiogram (ECG) systems, risk factors for sudden cardiac death in patients with malignant tachyarrhythmias and in myocardial infarction survivors, experimental and clinical studies on heart rate variability, and clinical studies on implantable cardioverter-defibrillators.

Antzelevitch, Charles. Dr. Antzelevitch was recruited by Dr. Gordon Moe as a postdoctoral fellow in 1977. Working under Emilio Kabela, MD, at the Upstate Medical Center Syracuse, N.Y., he rushed to complete his thesis work so as to be available in Utica, N.Y. by August 1977, when Moe was scheduled to return from Europe. He regarded his work with Moe and with Dr. "Pepe" Jalife, whom he had befriended while in Syracuse, as the opportunity of a lifetime. Antzelevitch initially joined Moe and Jalife to study parasystolic arrhythmias, and helped to further define the mechanisms by which naturally occurring or ectopic pacemakers can be modulated by the influences of surrounding tissues, a concept known as "modulated parasystole." Antzelevitch discovered a phenomenon called "electrotonic inhibition," which is a process whereby a subthreshold event can exert a negative effect on the excitability of cardiac tissues and thereby can influence conduction of an electrical impulse in the heart. He also initiated testing for the effectiveness of antiarrhythmic drugs using models of parasystole and reflected reentry. He is currently

Executive Director and Director of Research at the Masonic Medical Research Laboratory in Utica, New York.

Aristotle. (*384 BC, Stagira/Chaldice; †322 BC, near Chalcis/Euboea). Philosopher. From 367 BC until Plato's death (348–347 BC), Aristotle belonged to Plato's Academy. Later he founded the "Peripathetic School" in Athens, which was named for its location, the covered walkways in the Lyceum. In the history of medicine, his natural philosophical writings, including *De Anima* (On the Soul), *De Generatione Animalium* (On the Generation of Animals), and *De Partibus Animalium* (On the Parts of Animals), have gained great importance. Accordingly, Aristotelian physiology was based on the polarity of the heart (red, warm blood) and the brain (pale, cold slime), in which the heart was the central organ of the human body and therefore was regarded as the seat of the soul.

Aristotle

Asclepius. Legendary god of healing. Asclepius was the most important god of healing among the Greeks, and was regarded from the end of the fifth century BC as the forefather of the Hippocratic physicians, who are also called "Asclepiads" and "Asclepius' disciples." In Homer's *Iliad*, he is regarded as a hero and a physician. According to legend, he is the son of a mortal woman and Apollo, who was also venerated as a god of healing. Since the late sixth century, numerous shrines (Asclepieia) dedicated to Asclepius are noticeable in the Mediterranean region, in which a ritual healing cult was practiced until late antiquity. (See bust on p. 8.)

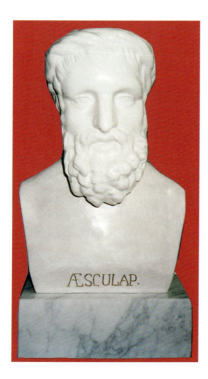

Asclepius

Auricchio, Angelo. Has been a leader in the development of non-pharmacological therapies for heart failure since its inception in the mid 1990s. Having been born in Italy, he went to medical school in Naples and worked in both Hannover and Rome before moving to the Otto von Guericke University in Magdeburg, Germany, in 1994. He is an Attending Physician and Associate Professor of Medicine in the Division of Cardiology at the University Hospital and Director of the Cardiac Catheterization Laboratory. In 1991, Auricchio was awarded the North American Society of Pacing and Electrophysiology Young Investigator Award. He is active on many committees, and although his main research interests are in heart failure therapies and in particular biventricular pacing, he is also interested in interventional cardiology and magnetic resonance imaging. Auricchio has published widely, having contributed to more than 50 original articles and book chapters and 180 published scientific abstracts.

Avicenna. Arabic name: Abû Alî al-Husain ibn Abd Allah ibn Sînâ al Qânûnî (*980, Afsana/Persia; †June, 1037, Hamadan). Physician and philosopher. In addition to his importance as a representative of Aristotelian philosophy, he led Arab medicine to its apex with his

Avicenna

(predominantly theoretical) main work, *Canon Medicinae* (Canon of Medicine), which he composed around 1030 while serving as a personal physician of the Caliph of Baghdad.

Bakken, Earl E. Earl Bakken was born in 1924 in Minneapolis, Minnesota, where he grew up with a strong Norwegian heritage. He served 3 years in World War II in the Army Signal Corps as a radar instructor. He returned to Minneapolis and earned his BSc degree and then his Masters degree in electrical engineering at the University of Minnesota. On April 29, 1949, Earl Bakken and his brother-in-law, Palmer Hermundslie, founded Medtronic Inc., and set up shop in a garage in northeast Minneapolis for the purpose of repairing medical equipment. On occasion, they also built devices on special order. One such order for a battery powered external pulse generator came from Dr. C. Walton Lillehei, who approached Bakken in October of 1957 and asked him to make a better pacemaker than the AC pacemakers then in use in intensive care units. Later, Medtronic became the first company to commercialize an implantable pacemaker. From 1960 the company has been the world leader in pacemakers in market share. Since his "retirement" from Medtronic, Bakken has founded and developed the Bakken Library and Museum, which showcases the role of electricity in medicine and life. Furthermore, he has been instrumental in the development of the North Hawaii Community and the Archeaus Project, which is working to develop Hawaii as a "healing island."

Barold, S. Serge. Born in Paris, France, in 1936, Dr. Barold survived childhood in war-torn Europe. Polylingual, brilliant, with a compulsion for excellence, he found his place in Australia. He graduated at Sydney University Medical School in 1960, at the top of his class, and won six awards for distinction. He was awarded fellowships in cardiology at the National Heart Hospital in London, UK, Mount Sinai Hospital in New York, and Mount Sinai Hospital in Miami, Florida. Finally he became an US citizen and, in 1969, he chose Rochester, New York, as his home. Rising through the ranks, he became Chief of Cardiology at Genesee Hospital and Professor of Medicine at the University of Rochester School of Medicine and Dentistry, where he now has emeritus status. Over the years, he has presented many important papers on hemodynamics, arrhythmias, and pacing. Barold now lives in Florida, and he is still very active, particularly in the fields of pathogenesis of atrioventricular (AV) conduction and history of cardiac arrhythmias. He is a member of the NASPExAM (North American Society of Pacing and ElectrophysiologyxAM) Writing Committee and an Associate Editor of *Pacing and Clinical Electrophysiology: PACE*. He was honored with the Distinguished Teacher Award from NASPE in 1992.

Bayés de Luna, Antonio. Dr. Bayés de Luna was born in Vic, near Barcelona, Spain, in 1936. He was trained as a cardiologist at the Institute of Cardiology at the Hammersmith Hospital in London, UK, and at the School of Cardiology at the University of Barcelona. He obtained his degree in cardiology in 1964. He is currently a Full Professor of Cardiology at the Autonomous University of Barcelona and Director of the Catalan Institute of Cardiology at Hospital Sant Pau. He has been President of the Catalan Society of Cardiology, the Spanish Society of Cardiology, and the World Heart Federation. He is now President of the Organizing Committee of the "World Heart Day." He has participated in numerous activities within the European Society of Cardiology (ESC). He has been a member of the Scientific Executive Committee, the Nominating Committee for the Nomination of the Board of ESC, the Library and History Committee, and the Executive Committee of the European Board of Specialists in Cardiology (EBSC). Bayés de Luna has been involved with a wide range of articles on medical care, teaching, and research. His book on electrocardiograms (ECGs) has been translated into seven languages.

Benditt, David Guay. Dr. Benditt was born in Winnipeg, Manitoba, Canada, on November 17, 1946. His present position is Professor of Medicine in the Cardiovascular Division of the Department of Medicine, as well as the Codirector of the Cardiac Arrhythmia Center at the University of Minnesota Medical School. Benditt received his training in cardiology/electrophysiology from 1975 to 1978 at the Duke University

Medical Center, Durham, North Carolina. He has received numerous honors and awards, and was President of the North American Society for Pacing and Electrophysiology 1993–1994. He is an active member of several important editorial boards, including *The American Journal of Cardiology*, *The Journal of Cardiovascular Electrophysiology*, *Europace*, *Journal of Interventional Cardiac Electrophysiology* (Associate Editor), and *Chinese Journal of Cardiac Arrhythmias*. Benditt has published many pioneering scientific articles in the leading pacing and electrophysiology journals. He is also a fellow of the American College of Cardiology and a fellow of the Royal College of Physicians of Canada.

Berkovits, Barouh. Between the late 1950s and the mid 1960s Berkovits worked as an engineer at the Medical Appliance Division of the American Optical Company, where he invented a heart monitor, the first closed-chest DC defibrillator, and the DC cardioverter. Berkovits then invented the "demand" or VVI pacemaker, initially as an external device, then as part of an implantable system.

Bigger, J. Thomas, Jr. Dr. Bigger was born and educated in the southern USA, attending Emory University and the Medical College of Georgia. He played a pioneering role in the establishment of patient risk after myocardial infarction by leading a series of critical prospective, randomized trials. This work provided pivotal clinical information that has defined the appropriate care for survivors of myocardial infarction. Bigger was the recipient of the 1998 Distinguished Scientist Award of the North American Society of Pacing and Electrophysiology.

Blanc, Jean-Jacques. Dr. Blanc was born in Paris, France, on August 1, 1945. In 1994, he became a Professor of Cardiology. His present position is Chief of the Department of Cardiology at the University Hospital of Brest, France, a position he has held since 1985. Blanc has been Chairman of the Working Group of Cardiac Stimulation of the French Society of Cardiology since January 1, 2000. That year he was elected as a board member of the French Society of Cardiology (*Société Française de Cardiologie*). He is member of the North American Society of Pacing and Electrophysiology and a fellow of the European Society of Cardiology. Blanc has published more than 100 articles in peer-reviewed national and international journals.

Bloch Thomsen, Poul Erik. Dr. Bloch Thomsen is Chairman of the Cardiology Department and Cardiology Research Laboratory at Gentofte Hospital, University of Copenhagen, Denmark. His specialties are cardiac electrophysiology, cardiac arrhythmias, and cardiac pacemakers, and he has contributed to the literature numerous notable papers and abstracts in these areas. He is one of the principal investigators as well as a member of the Steering Committee in the two DIAMOND studies. Bloch Thomsen was Chairman of the European Working Group of Cardiac Pacing and Arrhythmia (EUROPACE) 2001 in Copenhagen.

Blomström-Lundqvist, Carina. Dr. Blomström-Lundqvist was born in Stockholm on January 5, 1954. Her present address is Department of Cardiology, University Hospital in Uppsala, S-75185 Uppsala, Sweden. Blomström-Lundqvist received her PhD degree in 1987 with the thesis "Arrhythmogenic right ventricular dysplasia diagnostic and prognostic implications." In 1988, she became a Specialist in Cardiology; she was appointed as Associate Professor in Cardiology in 1990, and served as Scientific Secretary of the Swedish Society of Cardiology from 1993 to 1997. Blomström-Lundqvist was elected as General Secretary of the Swedish Medical Association 1999 (re-elected 2001). Her present appointment is Professor in Electrocardiology at University Hospital in Uppsala (since 2000); she is also Head of Section of Electrocardiology, Department of Cardiology, Uppsala University Hospital. In 2002, Blomström-Lundqvist was elected as Chairman of the Working Group of Arrhythmias of the European Society of Cardiology.

Boerhaave, Herman. (*December 31, 1668, Voorhout near Leiden; †September 23, 1738, Leiden). Clinician, botanist, and chemist. In 1714 he also became a Professor of Applied Medicine, and in 1718 a Professor of Chemistry. In his writings *Institutiones Medicae* (Principles of Theoretical Medicine; Leiden, 1708) and *Aphorismi de Cognoscendis et Curandis Morbis* (Aphorisms on Diseases to be Recognized and Healed; Leiden, 1709), he attempted, on the basis of Hippocratism, to synthesize the various medical trends of his time (iatrochemistry, iatrophysics, and vitalism). From 1714 he began conducting student teaching at the patient's sick bed at the Cäcilien Hospital in Leiden. For this reason, he is considered in the history of medicine the founder of modern clinical instruction delivered at the patient's sick bed (bedside teaching).

Borbola, Joseph. Dr. Borbola was born in Holzminden, Germany on September 10, 1945. He received his MD from the Albert Szent-Györgyi University of Szeged, Hungary, in 1970. He completed several fellowships in internal medicine and cardiology in Hungary; later on at the Medical Faculty of the University of Tübingen, Germany (1975); Harvard Medical School (mentor: Prof. Bernard Lown, Cardiovascular Laboratory) in Boston (1985–1986); Rush-Presbyterian University,

Chicago (1986–1987). Borbola has been working at the National Institute of Cardiology since 1975. He became Professor of Cardiology in 1998 and Head of the Department of Cardiology in 2003. His interests include diagnostics and treatment of cardiac arrhythmias, especially antiarrhythmic drugs, radiofrequency current ablation, pacemakers, and implantable cardioverter-defibrillator (ICDs). He became a Fellow of the European Society of Cardiology (ESC) in 1990. He served at the Board of the ESC as councilor from 1996 to 1998. He was awarded with the silver medal of the ESC (1998). Borbola currently is one of the Nucleus Members of the ESC Working Group on Cardiac Arrhythmias. He was a Board member of the International Society of Cardiac Pacing and Electrophysiology, Alpe–Adria Association of Cardiology, and the Hungarian Society of Cardiology. Borbola has published more than 160 scientific papers in national and international journals and has authored seven books.

Borggrefe, Martin. Dr. Borggrefe was born in 1955. His graduate thesis deals with the catheter ablation of tachycardic rhythm disturbances using high-frequency current in the course of both experimental and clinical studies. He is the principal investigator of the Catheter Ablation Registry, an international prospective registry for evaluating the safety and long-term effectiveness of catheter ablation. He also set up a European registry for recording complications in high-frequency catheter ablations. Borggrefe was country coordinator of the SWORD (survival with oral D-sotalol) study. He is well known for his work on high-frequency current ablation of an accessory pathway in humans. Since the summer of 2000 he has served as the Director of the II. Medical Clinic at the University of Heidelberg in the Mannheim Municipal Hospitals in Mannheim, Germany. He is also a fellow of the European Society of Cardiology.

Bouvrain, Yves. (*1910; †January 21, 2002). Dr Bouvrain was an Honorary Professor of Cardiology and Emeritus Member of the French Academy of Medicine. In 1938, Bouvrain described a form of skin lymphome known as "Lymphoma of Sézary–Bouvrain" (*Bulletin of the French Society of Dermatology*, February 1938). In 1961, he created the world's first intensive cardiac care unit. His experience with cardiac resuscitation at Hôpital Lariboisière in 1962 was published in *La Presse Médicale* in May 1963. From 1961 to 1964 he published many articles (mainly in French) on electrotherapy in ventricular tachyarrhythmias using external electric shock, and on suppression of supraventricular paroxysmal tachycardias by electrotherapeutic means. In 1961 (with F. I. Zacouto), he published a report describing a combination of devices that he called a "resuscitation device." This combination of devices consisted of an electrotherapeutic monitor, a defibrillator, and a pacemaker for antitachycardia pacing. Robert Slama, Philippe Coumel, and Jacques Mugica (his son-in-law) were his assistants.

Brachmann, Johannes. Dr. Brachmann was born in 1952. From 1980 to 1981 he was a Research Fellow of the German Research Society at the University of Oklahoma, in the field of experimental and clinical electrophysiology (Profs. Scherlag and Lazzara). In 1996, he became a Professor of Medicine, and in 1998 Chief Physician of Cardiology at the Coburg Hospital Coburg, Germany. His research interests are clinical electrophysiology (anti-arrhythmic agents, pacemakers, implantable cardioverter-defibrillators [ICDs]), and interventional cardiology (stents, atherectomy).

Bredikis, Jurgis. Dr. Bredikis began performing closed mitral surgery in Kaunas, Lithuania, in 1958, and soon followed with open-heart surgery. In 1959, he began transcutaneous cardiac pacing, and in 1961 he implanted myocardial electrodes with an external pulse generator. The following year he implanted a pulse generator, the first implant performed in Eastern Europe. In 1973, he became a founding member of the International Cardiac Pacing Society (the sponsor of the quadrennial World Symposia). By that time he had developed the largest service for pacemaker implantation, open-heart surgery, and arrhythmia surgery in the (former) Soviet Union. He performs these procedures at the Kaunas Medical Institute, where he has also sponsored numerous international conferences. In 1983, he performed the first laser ablation of an accessory pathway, and, over the years, has performed more than 1000 laser ablations for the Wolff–Parkinson–White syndrome. After retirement from active surgery, he served as the Lithuanian Ambassador to the Czech Republic.

Breithardt, Günter. Dr. Breithardt was born January 19, 1944 in Haan, Rhineland, Germany. He is currently Head of the University Hospital Münster—Internal Medicine C, and Chairman of the Division of Coronary Heart Disease of the Institute for Research of Arteriosclerosis at the University of Münster. He is a Professor of Medicine and Cardiology as well as a distinguished electrophysiologist. He and his group have long been interested in cardiac arrhythmias, including analysis and management and the use of implantable cardioverter-defibrillators. Breithardt is a former President of the European Society of Cardiology and of the German Society of Cardiology. He was also President of the Ninth World Symposium on Cardiac Pacing and Electrophysiology in Berlin. One of his main contributions is in the area of signal averaging.

Brignole, Michele. Dr. Brignole was born in Borzonasca, Italy, on August 17, 1952. He received his Degree in Medicine with excellence from the University of Genoa in 1976, and received his postgraduate training in Cardiology from the University of Pavia in 1986. He received his postgraduate training in Sport Medicine from the University of Genoa in 1990. From 1980 he practiced in the Division of Cardiology of the General Hospital of Lavagna, Genoa; from 1990 to 1996 he served as Director of the Section of Arrhythmology. His special interests include clinical diagnostic electrophysiology, catheter ablation therapy of tachyarrhythmias, pacing therapy, and implantation of automatic implantable cardioverter-defibrillators. In 1996, he became Chief of the Department of Cardiology. Brignole is a Fellow of the European Society of Cardiology (ESC). He currently is the Chairman of the Task Force on Syncope of the ESC. He serves as reviewer in many international journals; for example, *The Lancet, Circulation, Journal of the American College of Cardiology, The American Journal of Cardiology, Heart, Pacing and Clinical Electrophysiology: PACE, Stroke, European Heart Journal,* and *Europace,* where he is a member of the Editorial Board at the same time. Brignole's main research fields concern diagnosis, pathophysiology, and therapy of syncope, and rhythm disturbances and radiofrequency catheter ablation of arrhythmias. He has published more than 130 original papers in peer review journals (first author of 65 papers) and presented 55 invited articles and book chapters.

Brugada, Josep. Dr. Brugada was born on June 18, 1958 in Girona, Spain. In 1987, he obtained his doctoral degree from the University of Barcelona, and graduated *cum laude.* Between 1988 and 1991 he was investigator for the Dutch Royal Academy of Arts and Sciences, and a Professor of Physiology at the University of Limburg in Maastricht, the Netherlands. Since 1991 he has been the Director of the Arrhythmia Unit at the Clinic Hospital, Director of the Pediatric Arrhythmia Unit of the Sant Joan de Du Hospital, and an Associate Professor of Medicine, all at the University of Barcelona. In 1992, together with his brother Pedro, he described the so-called "Brugada syndrome." (See his article (1992) Right bundle branch block, persistent ST segment elevation and sudden cardiac death: A distinct clinical and electrocardiographic syndrome. *Journal of the American College of Cardiology,* **20,** 1391–6.) Their youngest brother, Ramon, is also active in this field. Brugada received the Award of the Fritz Acker Foundation of the German Society of Cardiology in 1998.

Brugada, Pedro. Dr. Brugada was born in Girona, Spain, on August 11, 1952. He was the Director of the Clinical Electrophysiology Laboratory

Pedro Brugada

at the University of Limburg, Annadal Hospital, in Maastricht, the Netherlands from 1982 to 1990. From 1988 to 1990 he was Head of the Coronary Disease Division at the Institute for Cardiovascular Diseases, University of Limburg, Maastricht. In 1989, he became Professor of Cardiology at the University of Limburg and Interuniversity Cardiology Institute in the Netherlands. Since 1991 he has been a staff member at the Cardiovascular Center, OLV Hospital in Aalst, Belgium. Brugada has received many honors and awards, including the Award of the Section of Electrocardiology of the Spanish Society of Cardiology (1979), the Award of the Fritz Acker Foundation from the German Society of Cardiology (1998), and the Third Mirowski Award, Madrid, Spain (1999). In 1999, he became an honorary member of the College of Physicians (Girona, Spain). He was elected as the 2001 Cardiologist of the Year in Paris, France. Brugada is also an honorary member of many international societies. He serves on several editorial boards, including *Pacing and Clinical Electrophysiology: PACE, Journal of Electrophysiology, New Trends in Arrhythmias,* and *Europace.* Brugada has published more than 800 scientific articles on electrophysiology and pacing. Not long ago, he and his brother Josep described a syndrome that is named after them "the Brugada syndrome."

Brugada, Ramon. Dr. Brugada was born in Girona, Spain, on September 1, 1966. He received his Doctor of Medicine degree in 1990 at the Autonomous University of Barcelona. His is currently Assistant Professor of Medicine, Section of Cardiology, at Baylor College of

Medicine in Houston, Texas. He has received several honors and awards, including the Young Investigator Award of the American College of Cardiology (1997), the Award of the Fritz Acker Foundation of the German Society of Cardiology (1998), the Josep Trueta Award of the Academia Ciencies Mediques Catalunya I Balears, Catalonia, Spain (1999), and the 1999 Mirowski Award on the occasion of the Arrhythmia Meeting held in Madrid, Spain. Brugada has editorial responsibilities to several scientific journals, including *Current Cardiology Reports*, *Timely Topics in Medicine*, *Investigacion Cardiovascular*, and *Cardiologie al Dia*. He is Genetics Coordinator in the Working Group on Brugada Syndrome, and a member of the Board of Directors of the American Institute for Catalan Studies.

Cain, Michael E. Dr. Cain was born on November 15, 1949 in Philadelphia, Pennsylvania. His present position is Tobias and Hortense Lewin Professor of Medicine and Director of the Cardiovascular Division at Washington University School of Medicine in St. Louis, Missouri. Cain was educated at Gettysburg College in Gettysburg, Pennsylvania. He earned his MD degree at the George Washington School of Medicine in Washington, D.C. His postgraduate training includes an internship and residency in medicine at the Washington University School of Medicine/Barnes Hospital in St. Louis, Missouri (1975–1977), research and clinical cardiology fellowship at the Washington University School of Medicine/Barnes Hospital (1977–1980), and clinical electrophysiology fellowship at the University of Pennsylvania School of Medicine in Philadelphia. Cain has received numerous honors and awards, including the Arthur E. Strauss Award from the American Heart Association (2000). He has editorial responsibilities to several journals, including *Circulation*, *The American Journal of Cardiology*, *Journal of the American College of Cardiology*, and *The Journal of Cardiovascular Electrophysiology*. Currently, he is principle investigator of the National Institutes of Health (NIH) supported project, "Pathophysiologic Basis of Ventricular Tachycardia in Man." Cain's bibliography lists 127 manuscripts including numerous highly ranking publications in major peer-reviewed journals.

Calkins, Hugh Grosvenor. Dr. Calkins was born in Boston, Massachusetts on December 20, 1956. After internship and residency at the Massachusetts General Hospital in Boston, he was post-doctoral fellow at the Division of Cardiology, The Johns Hopkins Hospital (Dr. Myron L. Weisfeldt, Chief of Cardiology) from 1986 to 1989. His current appointments are Professor of Medicine, Department of Medicine, Division of Cardiology, and Professor of Pediatrics, Department of Pediatrics, Division of Pediatric Cardiology, The Johns Hopkins

University School of Medicine, and Director of the Arrhythmia Service, Clinical Electrophysiology Laboratory, Tilt Table Diagnostic Laboratory and Arrhythmogenic Right Ventricular Dysplasia Program, The Johns Hopkins Hospital, Baltimore. His research activities include a broad spectrum of cardiac arrhythmias, particularly electric pacemakers and automatic implantable cardioverter-defibrillators. Calkins received numerous awards and honors, including the Helen B. Taussig Award of the American Heart Association–Maryland Affiliate, Inc. (2000), Visiting Professor, Cleveland Clinic Foundation (2001), Best Doctors Award (2002), and America's Top Physicians Award (2003). He has two children, Emily Nichols C., Feb. 21, 1992, and Eliza Nichols C., Nov. 15, 1995.

Camm, Alan John. Dr. Camm was born on January 11, 1947 in Lincolnshire, UK, and obtained his baccalaureate degree, with a major in physiology, at the University of London, and his medical education at Guy's Hospital Medical School, also at University of London, UK. His doctoral thesis, entitled "The application of pacemakers to tachycardia termination," was accepted by the University of London in 1981. He joined the faculty at St. Bartholomew's Hospital in London and was appointed the Sir Ronald Bodley Scott Professor of Cardiovascular Medicine in 1983. In 1986, he became Professor of Clinical Cardiology and Chairman at the St. George's Hospital Medical School in London.

A. John Camm

Since then he has held the Chair in Internal Medicine, as well as numerous appointments in British, European, and international medical organizations. He was the first International Trustee of the North American Society of Pacing and Electrophysiology, and is currently Honorary Physician at the Court of St. James and President of the British Cardiac Society. He holds editorships at *Clinical Cardiology, Pacing and Clinical Electrophysiology: PACE*, and the *Journal of Interventional Cardiac Electrophysiology*. Camm is a prolific researcher, having published more than 680 articles, 970 research abstracts, and 148 chapters in various books. He has received numerous honors and awards as well as honorary memberships. As a scientist, he holds many important offices and memberships on the editorial boards of several leading publications. In 2001, the North American Society of Pacing and Electrophysiology awarded him the Distinguished Teacher Award. He is coeditor of a major textbook entitled *Electrophysiological Disorders of the Heart* (see historical page 208).

Cammilli, Leonardo. Dr. Cammilli began his involvement in cardiac pacing with the design and implantation of a radiofrequency pacemaker in the early 1960s. In 1976, he devised and implanted the first rate-modulated pacemaker using a pH sensor to drive the cardiac rate, thus beginning the era of rate-modulated pacing. The device functioned for several years in the patient. In the 1990s Cammilli developed a drug delivery system that discharged into the coronary sinus as a ventricular defibrillator. He performed the first human pH-triggered, rate-modulated pacemaker implant on January 5, 1977, and is therefore the father of this critical aspect of cardiac pacing.

Campbell, Ronald W.F. (*1946, Scotland; †June, 1998, Spain). Dr. Campbell attended medical school in Edinburgh, Scotland, and trained as a research fellow in cardiac arrhythmias at Duke University, Durham, North Carolina. Upon his return to the UK, he moved to the Freeman Hospital in Newcastle-upon-Tyne, where he remained for the rest of his career. Campbell rose to become a British Heart Foundation Professor of Clinical Cardiology and the Chief of the Academic Cardiology Department. At the time of his death, he was President of the British Cardiac Society. His research and publications dealt with ventricular arrhythmias and the effect of antiarrhythmic medications, atrial fibrillation, and its association with accessory pathways and sudden death. Later in life, his research efforts extended to the mechanisms of ventricular arrhythmias and myocardial infarction; most recently, his interest focused on non-invasive means of predicting increased risk of sudden cardiac death. He was a master lecturer in the English language. He was a smiling, happy person with an infectious sense of humor. In a

Ronald W.F. Campbell

review of the book, "History of the disorders of Cardiac Rhythm" (1995), Campbell wrote: ". . . perhaps my name will feature in future editions, but I think not. This book is devoted to the 'greats' in the field of arrhythmology." This has become true in a very unfortunate, unexpected way, and it should be emphasized that he truly was one of the "greats."

Cannom, David S. A Clinical Professor of Medicine at the University of California, Los Angeles School of Medicine, Dr. Cannom also serves as Director of Cardiology at the Good Samaritan Hospital in Los Angeles, California. He is also a managing partner of an 18-member cardiology group, the Los Angeles Cardiology Associates, which specializes in coronary and electrophysiological interventions. He received his degree in medicine from the University of Minnesota Medical School, completed his medical training as an internship and residency at Yale–New Haven Hospital in New Haven, Connecticut, and his cardiology training at Stanford University. Cannom is past governor and President of the California chapter of the American College of Cardiology. He is also past President of the North American Society of Pacing and Electrophysiology. He serves on the editorial boards of *Pacing and Clinical Electrophysiology: PACE, Cardiac Electrophysiology Review*, and the *Journal of Cardiac Arrhythmias Index and Reviews.*

Cappato, Riccardo. The cardiologist Dr. Cappato was born on May 2, 1958 in Ferrara, Italy. His current professional address is Arrhythmia and Electrophysiology Department, Policlinico San Donato, Via Morandi 30, I-20097 San Donato Milanese, Milan, Italy. Cappato's medical experience includes medical studies at the Università degli Studi di Ferrara (1977–1983), internship at the Institute of Internal Medicine at the Università degli Studi di Ferrara (1983–1984), and residency at the Divisione di Cardiologia of the Arcispedale S. Anna di Ferrara (1984–1989). Cappato became a junior physician at the Department of Cardiology of the Arcispedale S. Anna di Ferrara (1989–1993) and Associate Researcher and Codirector of the Electrophysiology Laboratory at the Department of Cardiology, University of Hamburg (Germany) (1993–1995), and later on faculty member and Codirector of the Electrophysiology Laboratory at the Department of Medicine II., St. Georg Hospital Hamburg (Germany), headed by Prof. Karl-Heinz Kuck. In 2000, Cappato became Chief, Dipartimento di Aritmologia Clinica ed Elettrofisiologia at the Policlinico San Donato, Milan. Cappato is not married.

Capucci, Alessandro. Dr. Capucci, born November 7, 1948, studied medicine at the University of Bologna, Italy, and graduated in 1973. His qualifications include specializations in cardiology (1976), internal medicine (1978), and sports medicine (1984). In 1986, he was appointed Chef de clinique at the Institute of Cardiology of the University of Bologna, where he was in charge of the direction of the Laboratory of Electrophysiology and Pacemaker implants and the Center of the Study of Arrhythmias at the Institute of Cardiology. Capucci now heads the Cardiology Division at Ospedale Civile in Piacenza, Italy.

Castellanos, Agustin. Dr. Castellanos was born in Havana, Cuba, on September 12, 1929. In June 1945, he received his BSc degree from the Baldor Institute in Havana, and subsequently attended the University of Havana School of Medicine, where he received his Doctor of Medicine degree in 1953. His specific interest in arrhythmias and conduction disturbances was apparent in his early publications, in which he addressed such topics as bundle branch block, aberrant intraventricular conduction, supraventricular tachycardias with block, applications of vectorcardiography, and the effect of new antiarrhythmic agents on various tachycardias. Castellanos' internationally respected, most important publications focus on the concepts of hemiblock, the evolution of pacemakers, applications of vectorcardiography, the understanding of accessory pathways and other patterns of intraventricular conduction abnormalities, and the use of invasive electrophysiological testing. He received the Distinguished Scientist Award of the North American Society of Pacing and Electrophysiology in 1990 (see historical page 194).

Chardack, William. From 1958 to 1960 Dr. Chardack worked at the Veterans Affairs Hospital in Buffalo, New York, where he collaborated with engineer Wilson Greatbatch on designing and clinically testing what was at that time the most advanced implantable pacing system. Chardack later invented highly reliable myocardial and endocardial leads based on a coiled-spring design, and contributed many other innovations to the field of cardiac electrostimulation.

Chen, Shih-Ann. Dr. Chen was born December 22, 1959 in Chaii, Taiwan. His present position is Professor of Medicine at National Yang-Ming University, School of Medicine and Director of Cardiac Electrophysiology Laboratory, Division of Cardiology, Department of Medicine at the Veterans General Hospital—Taipei, in Taiwan. Chen has received numerous honors and awards, including two Outstanding Research Awards from the Society of Cardiology, Taiwan (1990, 1993), and an Outstanding Research Award from the Min–Tsai Medical Foundation (1993). He serves on several scientific/editorial boards, including the *Journal of Interventional Cardiac Electrophysiology, Journal of Cardiovascular Electrophysiology*, and *Pacing and Clinical Electrophysiology: PACE.* He has published pioneering papers on electrophysiology, particularly therapeutic ablation procedures.

Clémenty, Jacques. Dr. Clémenty was born in Thézan-les-Béziers, France, on January 5, 1945. In 1975, he received his MD degree and became Head of the Cardiac Clinic of the Hospitals of Bordeaux, France. Clémenty is a founding member of the Working Group of Hypertension of the French Society of Cardiology. He is also a member of the French Society of Cardiology and a member of the Working Group on Cardiac Stimulation of the French Society of Cardiology. Clémenty is one of the leading educators in France. He has published 700 scientific articles dealing specifically with electrophysiology, hypertension, and cardiovascular pathology.

Cobbe, Stuart Malcom. Dr. Cobbe was born in 1948. He is currently Walton Professor of Medical Cardiology, University of Glasgow, and Head of the Department of Medical Cardiology at the Royal Infirmary in Glasgow, Scotland. His degrees and qualifications include his MD thesis, accepted in 1981 by the University of Cambridge, and Fellow of the Royal College of Physicians (London and Glasgow) in 1986. Cobbe's research interests include cardiac metabolism, basic cardiac electrophysiology, pharmacology of antiarrhythmic drugs, cardiac arrhythmias and electrophysiology, myocardial infarction, coronary prevention, cardiopulmonary resuscitation, cardiac failure, and left ventricular hypertrophy.

Connolly, Stuart J. Dr. Connolly was born in Montreal, Canada, in 1949. He received his cardiology training at the University of Toronto and his postgraduate training in electrophysiology at Stanford University. In 1983, he joined the faculty of McMaster University, where he is now a Professor and the Director of the Arrhythmia Service and the Electrophysiology Laboratory. His main research interests have been in the area of randomized clinical trials, and he has been involved in the design and implementation of a number of studies related to the care of patients with cardiac arrhythmias. Among the studies of which he was the principal investigator are the Canadian Atrial Fibrillation Anticoagulation (CAFA) study, the Canadian Amiodarone Myocardial Infarction Arrhythmia Trial (CAMIAT), the Canadian Implantable Defibrillator Study (CIDS), the Canadian Trial of Physiologic Pacing (CTOPP), and two Vasovagal Pacemaker Studies (VPS I and VPS-2).

Cosio, Francisco Garcia. Dr. Cosio was born in Oviedo, Spain, on December 21, 1943. From 1967 to 1971 he served his internship and residency at the Mount Sinai Hospital and Hennepin County General Hospital in Minneapolis, Minnesota. From 1971 to 1973 he served a cardiology fellowship at the University of Minnesota Hospitals in Minneapolis. His present position is Chief of the Cardiology Service, Hospital Universitario de Getafe in Madrid, Spain. Cosio is a fellow of the American College of Cardiology, of the European Society of Cardiology (ESC), and of the North American Society of Pacing and Electrophysiology, as well as a member of the Nucleus of the Working Group of Arrhythmias of the ESC. He has worldwide editorial responsibilities, and has published pioneering papers in the most relevant peer-reviewed cardiology journals.

Coumel, Philippe. (*1935, Lyon, France; †March 18, 2004). Dr. Coumel was Head of the Cardiology Department at Hôpital Lariboisière, Paris, France, and spent more than 30 years working in the field of cardiac electrophysiology and clinical arrhythmias. He pioneered the exploration of arrhythmia mechanisms using invasive approaches of endocavitary recordings and programmed stimulation in junctional reciprocating tachycardias as well as atrial and ventricular tachyarrhythmias. Using the Holter technique, he demonstrated the importance of the role of the autonomic nervous system and rate-dependence in the determination of arrhythmias. Many of his publications focus on non-invasive electrophysiology with invasive investigations. He was founding fellow and former Chairman of the Working Group on Cardiac Arrhythmias of the European Society of Cardiology (see historical pages 188 and 211).

Cox, James L. Dr. Cox trained as a surgeon at Duke University Medical Center, turning his attention to cardiovascular research in his capacity as a fellow. He holds three US patents pertaining to cardiac surgery. He invented innovative antiarrhythmic surgical procedures, most notably the "Cox procedure" for atrial fibrillation. Cox is renowned for his outstanding antiarrhythmic cardiosurgical achievements and, in 1996, was granted the Distinguished Scientist Award by the North American Society of Pacing and Electrophysiology.

Cranefield, Paul F. Dr. Cranefield was born in Madison, Wisconsin, on April 28, 1925. He graduated from the University of Wisconsin in 1946 with a PhD in mathematics and obtained his PhD degree in physiology in 1951 from the same institution. Cranefield's work has been seminal in the development of several fields of clinical cardiac electrophysiology. This includes his fundamental work on atrioventricular (AV) nodal conduction, his work on excitation in depressed tissues and the slow response, his pioneering work in the identification and characterization of triggered rhythms (both early and delayed after depolarizations), and his work on reentry.

Crijns, Harry. Dr. Crijns was born on August 9, 1954 in Voerendaal, the Netherlands. He graduated from the Medical School of the University of Amsterdam in 1981. He served a fellowship in experimental electrophysiology from 1978 to 1979 in the Department of Cardiology at the Wilhelmina Gasthuis in Amsterdam, under the supervision of Prof. M. J. Janse and the late Prof. D. Durrer. In 1993, he defended his thesis entitled "Changes of intracardiac conduction induced by antiarrhythmic drugs. Importance of use and reverse use dependence." In 1995, he was appointed Acting Chief of Cardiology. In 1997, he became Chair of the Department of Cardiology at the Thorax Center of the University Hospital Groningen. In 2001, he moved to Maastricht to be Chair of the Department of Cardiology of the University Hospital Maastricht. His scientific merits include membership in the Nucleus of the Working Group of Arrhythmias of the European Society of Cardiology and the North American Society of Pacing and Electrophysiology. His research concentrates on atrial fibrillation. He has published more than 100 scientific articles in peer-reviewed journals.

Curtis, Anne B. Dr. Curtis was born on April 16, 1953 in Brooklyn, New York. Her current professional address is Division of Cardiology, Department of Medicine, University of Florida, Health Science Center, 1600 S.W. Archer Road, Gainesville, FL 32610-0277. In 1996, Dr. Curtis became a consultant at the Veterans Affairs Medical Center, Gainesville,

Florida. From 1997 onwards she served as Professor of Medicine at the University of Florida, Gainesville, Florida. Her postdoctoral training includes a residency in internal medicine at the Presbyterian Hospital New York (1979–1982) and a fellowship in cardiovascular diseases, and later on clinical cardiac electrophysiology at the Duke University Medical Center, Durham, North Carolina (1982–1986). She received the American Board Certification of Internal Medicine in 2002. Previous university appointments include an Associate Professorship of Medicine (Director, Clinical Electrophysiology) at the University of Florida, Gainesville (1992–1997). Curtis received several honors and awards including *summa cum laude* with High Honors in Chemistry, Rutgers University (1975). She is married to Alexander Domijan, Jr., PhD, and has three children.

Damato, Anthony N. (*January 10, 1930, Jersey City, New Jersey, USA; †2001). In 1969, Dr. Damato, along with Drs. Kosowsky and Scherlag, described for the first time a reliable and easy technique using catheter electrodes to record a His bundle potential during a cardiac catheterization procedure. From this sprang the field of clinical cardiac electrophysiology. Initially, Damato and his colleagues became intimately involved not only in the demonstration of the His bundle potential, but also in understanding the nature of atrioventricular (AV) conduction, various forms of AV conduction block, and bradycardias. Subsequently, the technology grew to include programmed electrophysiological stimulation and the study of tachyarrhythmias. Those who trained in his laboratory included Drs. Akhtar, Berkovits, Cannom, Gallagher, Goldreyer, Josephson, Mirowski, Prystowsky, Rosen, Ruskin, Vardas, and Wit. In 1984, Damato became Director of the Department of Medicine at Jersey City Medical Center. He retired in New Jersey.

Daubert, Jean Claude. Dr. Daubert is a Professor of Cardiology at the University of Rennes and Chief of the Department of Cardiovascular Diseases at the University Hospital of Rennes, France. Daubert's specialty is clinical and technological research on cardiac pacing and arrhythmias. He is Chairman of the MUSTIC Study Board, a study group of the European Society of Cardiology which is working to assess the clinical efficacy of multisite pacing in different pathologies, especially refractory heart failure. Daubert gave the magisterial lecture on the topic of treating heart failure through the use of pacing devices, entitled "From ICD to cardiac resynchronization," in Rome on December 5, 2000.

Della Bella, Paolo. Dr. Della Bella was born on November 14, 1954 in Milan, Italy. His current professional position is Director of the

Arrhythmic Unit and of the Electrophysiology Laboratory of the Centro Cardiologico "Fondazione I. Monzion," Milan, Italy. He graduated in medicine and surgery at the University of Milan *cum laude* in 1979, and became a specialist in anesthesiology at the Second School of Specialization in Milan *cum laude* in 1990. From 1979 to 1981 he was resident at the Institute of Medical Pathology I. of Milan, directed by Prof. Zanchetti. In 1986, he joined the Laboratory of Clinical Arrhythmology of the University of Maastricht, the Netherlands, directed by Prof. H. J. J. Wellens, and studied the invasive techniques on the mechanism of cardiac arrhythmias and participated in the program of surgery of arrhythmias in patients with Wolff–Parkinson–White syndrome (Prof. Penn). In 1988, he spent some months at the Department of Cardiosurgery of the University of Western Ontario in London, Ontario, Canada, directed by Prof. Guiraudon, and spent 3 months at the Department of Electrophysiology in the Unit of Cardiology of the University of Oklahoma directed by Prof. Jackman (studies on catheter ablation techniques). In 1994, Della Bella became Director of the Department of Clinical Electrophysiology of the Centro Cardiologico, Milan. Since 1994, he has been a university professor at the School of Specialization in Cardiology of Milan. In 1995, he served on the editorial board of *Cardiologia*, and, in 1998, he became a member of the editorial board of *Europace*.

Descartes, René. (*March 31, 1596, La Haye, France; †February 11, 1650, Stockholm, Sweden). "Amateur physician," philosopher, mathematician, and founder of Cartesianism. Although Descartes had neither studied medicine nor practiced medicine, he devoted himself intensively

to anatomy and physiology, and outlined in his posthumously published work *De Homine* (1662; Fr. *Traité de l'homme*, 1667) a new theory of the human organism based on mechanics. He introduced the dualism of the body (*res extens*) and the soul (*res cogitans*) into medical teaching and declared the pineal gland as the seat of the soul. *"To investigate the truth, method is necessary."*

Dessertenne, François. Dessertenne was born in 1917. In 1943, he became an intern, and in 1948 an Assistant Professor of Medicine. He worked at Hôpital Lariboisière in Paris, France, and was the assistant of Prof. Bouvrain and later of Prof. Slama. In 1996, in his pioneering work, he described a variety of ventricular tachycardia characterized by two variable foci, which explained the twisted peaks. This particular aspect is now described using the French term "torsade de pointes." Dessertenne clearly made the distinction between torsade de pointes and ventricular tachycardia, although torsade de pointes may turn into ventricular tachycardia (see historical page 175).

Dodinot, Bernard. An early contributor to the development of atrial synchronous pacing during the 1960s, Dr. Dodinot was also Secretary General of the Third World Symposium on Cardiac Pacing in 1970. He has been involved in cardiac pacing since 1963 and is the founder and Editor of the French pacemaker journal *Stimucoeur*.

Doppler, Christian. (*November 29, 1803, Salzburg, Austria; †March 17, 1853, Venice, Italy). Doppler became a Professor of Mathematics and Physics in Prague and first presented his historic paper "Über das farbige Licht der Doppelsterne und einiger anderer Gestirne des Himmels,"

a completely theoretical paper (published by B & A, in Abhandlungen der Königlich Böhmischen Gesellschaft der Wissenschaften 1843; **2**, 465–82) on May 25, 1842 at a meeting of the Bohemian Scientific Society. The first attempts to verify his theory took place in Utrecht in February 1845, using a trombone orchestra that played in a railway station while a train passed. This had to be repeated in June because the first attempt was unsuccessful due to the winter weather conditions. In 1849 Doppler took over the position of Director of the newly founded Polytechnic Institute in Vienna. His invention of the "Doppler effect" revolutionized cardiology.

Dorian, Paul. Dr. Dorian is a Professor of the Faculty of Medicine and an Associate Member of the Graduate Faculty in the Department of Pharmacology at the University of Toronto in Canada. He received his MD degree in 1976 from McGill University in Canada. He is currently the Director of the Cardiac Electrophysiology Program at St. Michael's Hospital and Chairman of the Clinical Teachers' Association of Toronto. His research interests include cardiac electrophysiology and antiarrhythmic drug pharmacology. Dorian was previously the Editor-in-chief of *Clinics and Cases for Arrhythmia* and, over the years, has been a reviewer for major journals including the *Canadian Journal of Cardiology* and *Circulation*. His extensive publication record includes many articles and abstracts in the field of cardiology.

Dreifus, Leonard S. Dr. Dreifus was born on May 27, 1924 in Philadelphia, Pennsylvania. He received his medical training at Hahnemann Medical College. His present position is Clinical Professor of Medicine at the University of South Florida. Dreifus has received many honors and awards. His scientific merits include his work as President of the American College of Cardiology (1978–1979) and Director of the Division of Cardiovascular Diseases at Hahnemann University. His many honors include AOA (Alpha Omega Alpha), Distinguished Fellow of the American College of Cardiology, North American Society of Pacing and Electrophysiology Distinguished Teaching Award, and the Master Teacher Award, American College of Cardiology. To date, Dreifus has published 249 scientific articles in peer-reviewed journals.

Durrer, Dirk. (*1918; †1984). Dr. Durrer was a Professor of Cardiology and Clinical Physiology at the University of Amsterdam from 1957 to 1984. Based on T. W. Engelmann's "method of extrasystoles," Durrer introduced "programmed electrical stimulation" to gain insight into the mechanism of arrhythmias. In 1967, Durrer and his associates published their epochal papers on patients with Wolff–Parkinson–White syndrome. These studies gave solid support to the "reentry (circus movement) theory" as an explanation for certain types of tachycardia. The search also moved towards an understanding of ventricular tachycardias in chronic myocardial infarction Durrer's paper "Total excitation of the isolated human heart" (1970) provides the basis for understanding the genesis of the electrocardiogram (ECG). Durrer became an internationally recognized authority in the field of electrocardiology (see historical page 164).

Ector, Hugo. Dr. Ector was born in Elsene, Belgium, on August 19, 1943. He holds leading positions at the Department of Cardiology of the University Hospital Gasthuisberg, University of Leuven, and at the Faculty of Medicine of the Catholic University of Leuven, Belgium. Ector received his PhD in cardiology with his thesis on permanent cardiac pacing. He is the founder of the Belgian Working Group on Cardiac Pacing and Electrophysiology and President as well as a former Chairman of the Belgian Working Group on Cardiac Pacing, a working group of the European Society of Cardiology. In addition, he was organizer and Chairman of the Sixth European Symposium on Cardiac Pacing in 1993. Ector serves on several editorial/scientific boards, including *Stimucoeur* and *Pacing and Clinical Electrophysiology: PACE*. His primary area of interest concerns pacemaker leads. He has organized and coorganized numerous national and international symposia dealing with pacing leads and other electrophysiological topics.

Effert, Sven. (*March 31, 1922, Aachen, Germany; †January 9, 2000, Aachen). Dr. Effert studied medicine from 1940 to 1947 at the Universities of Bonn, Freiburg, and Düsseldorf, Germany. His studies were interrupted by military service during which he was wounded and captured as a prisoner of war. After his final medical examination and approval of his MD thesis in 1948, periods of professional training in pathological anatomy and internal medicine followed at the University of Düsseldorf. Effert was appointed Associate Professor in 1965. With the establishment of a medical faculty at the Technical University of Aachen in 1966, for the purpose of utilizing the engineering sciences for clinical medicine, Effert was appointed Professor of the Department of Internal Medicine I. Pacemaker technology, echocardiography, and invasive cardiological procedures were his key scientific activities. In 1959, he succeeded for the first time in producing echocardiographic documentation of an atrial tumor.

Einthoven, Willem. (*1860, Semarang, Java, Dutch East Indies [now Indonesia]; †1927). Einthoven is best known for applying the fundamental principles of physics to solving physiological problems. He was Professor of Physiology at the University of Leiden. On April 11, 1892,

using the Lippmann capillary electrometer, Einthoven recorded the first human electrocardiogram (ECG), following A. D. Waller's "A demonstration on man of electromotive changes accompanying the heart's beat" in England (1887/1889). Einthoven introduced the term "electrocardiogram" and published a paper in 1895 entitled "On the shape of the human electrocardiogram." Therein he used the characters P, Q, R, S, and T, for the first time, to indicate the deflections of the corrected tracing, a notation system that is still in use today. Starting from Deprez-d'Arsonval's moving-coil galvanometer, he finally developed his string galvanometer: "*un nouveau galvanometer*" (1901). With this new instrument, Einthoven made the first direct recording of the true human ECG in 1902. On October 23, 1924, Einthoven was awarded the Nobel Prize for Physiology and Medicine for his discovery of the mechanism of the ECG (see historical page 149).

Elmqvist, Rune. (*December 1, 1906, Lund, Sweden; †December 15, 1996). Dr. Elmqvist obtained his license to practice medicine in 1939. His main interests, however, were engineering and electromedicine. As a student he had already laid the foundation for the modern high-fidelity electrocardiogram (ECG) with his development of the band galvanometer and portable multichannel ECGs. In 1939, he joined the company that later became Siemens–Elema. Elmqvist invented and perfected ink-jet recording. The first ink-jet electrograph or mingograph was presented in 1950 at the first meeting of the European Society of Cardiology in Paris. Mingography rapidly became the gold standard for serious ECG work. Elmqvist was a physician who became an engineer and, consequently, the designer of much modern medical equipment. On October 8, 1958, a rechargeable, implantable cardiac pacemaker that he had designed and built was implanted. Even though it failed, it was the first pacemaker ever fully implanted. The first successful implantations did occur in London, UK, and Montevideo, Uruguay in 1960 (see historical page 173).

El-Sherif, Nabil. Dr. El-Sherif was born in Cairo, Egypt, on April 15, 1938. He graduated from Cairo University Medical School in December 1959, did his residency training at Cairo University Hospitals from 1960 to 1963, and joined the faculty of the Cardiology Department at Cairo University, where he remained from 1963 to 1972. He held an electrophysiology fellowship at Mount Sinai Medical Center in Miami Beach, Florida from 1972 to 1974, and then joined the University of Miami, where he was an Assistant Professor from 1974 to 1976 and an Associate Professor from 1976 to 1978. He became Professor of Medicine at the State University of New York, Downstate Medical Center and Director of the Cardiology Division at the Brooklyn Veterans Affairs Medical Center in 1978, and a Professor of Physiology in 1983. He was the Director of the Cardiology Regional Program at SUNY Downstate from 1984 to 1986 and has been Associate Chairman of Medicine for Research since 1986. El-Sherif is an internationally known cardiac electrophysiologist, clinician, and scientist. His major interest is the electrophysiological mechanisms of cardiac arrhythmias, especially those associated with myocardial ischemia/infarction. His work spans the whole field, from clinical investigations to animal models, ion channel physiology, and molecular biology studies. He has published more than 500 original papers, book chapters, and review articles, and has edited eight books. He is a member of several editorial boards and medical societies in the USA and abroad. He is an honorary member of the Egyptian Society of Pacing and Electrophysiology and the Albanian Society of Cardiology, and is Copresident of the Mediterranean Society of Pacing and Eletrophysiology and Coeditor of its research journal.

Engelmann, Theodor Wilhelm. (*1843; †1909). Dr. Engelmann was Professor of Histology and General Biology at the University of Utrecht, the Netherlands, between 1871 and 1888, Professor of Physiology at the University of Utrecht from 1888 to 1897, and Professor of Physiology at the University of Berlin, Germany, from 1897 to 1908. In his famous series "Observations and experiments of the suspended heart," Engelmann reported on his observations and recordings of a beating frog heart. In his writings, we, for the first time, come across the term "extrasystole" (both atrial and ventricular) and the observation that the compensatory pause after an extrasystole is lengthened exactly to the same degree as the interval preceding it is shortened. His most distinguished pupil and friend was Karel Frederik Wenckebach, who in 1903 dedicated his first book to his former teacher.

Epstein, Andrew E. Dr. Epstein was born on November 30, 1950 in New York. His current professional position and address is Professor of

Medicine, Division of Cardiovascular Disease, Department of Medicine, University of Alabama at Birmingham, Birmingham, Alabama 35294. His postgraduate hospital training includes internship and residency in internal medicine at Barnes Hospital, Washington University St. Louis, Missouri (Chief: Dr. David Kipnis) (1977–1978), fellowship and chief fellowship in cardiology (1980–1983) at the University of Alabama at Birmingham (Chiefs: Drs. Thomas N. James and Lloyd L. Hefner). Epstein's areas of research interest include defibrillation and cardioversion; electrophysiology and treatment of atrial and ventricular arrhythmias; and clinical trials in arrhythmia treatment. He serves in the editorial board of six key journals of the field, and is manuscript reviewer of the leading journals of cardiology. Epstein holds two patents and has presented frequently at national and international meetings. He is married to Eileen Marie Epstein and has one child (Anne Elizabeth Epstein).

Escher, Doris J.W. Dr. Escher founded a cardiac catheterization laboratory in 1948, participated in the development of transvenous pacing, and began the first pacemaker follow-up clinic. She has been involved in cardiac pacing from its inception, was coauthor of a 1970 book on cardiac pacing, has received the Pioneers in Cardiac Pacing and Electrophysiology and the President's Award from the North American Society of Pacing and Electrophysiology (NASPE), and has been President of NASPE.

Estes, N.A. Mark, III. Dr. Estes was born on August 20, 1949 in Newport, Rhode Island. He received his undergraduate degree with distinction from the University of Pennsylvania and his medical degree from the University of Cincinnati. After completing his medical and cardiology training in Boston, he trained in clinical cardiac electrophysiology at Massachusetts General Hospital. He has subsequently served as Chief of the Cardiac Arrhythmia Service at Tufts New England Medical Center and as Professor of Medicine at Tufts University School of Medicine. His research interests are in prediction and prevention of sudden cardiac death and cardiovascular disease in the athlete. He has published more than 200 manuscripts, 35 book chapters, and two books. He serves as a trustee for the North American Society of Pacing and Electrophysiology, Lifespan Healthcare System, Moses Brown School, the American Heart Association (AHA) New England Affiliate, and the Whitehead Institute of Massachusetts Institute of Technology. He has served as President of the New England Electrophysiology Society, and of the Boston board of the AHA, and is President of the New England Affiliate of the AHA.

Falk, Rodney. Dr. Falk is currently the Director of Clinical Cardiac Research and a Staff Cardiologist at Boston University Medical Center. He is also a Professor of Medicine at Boston University Medical School. He has a special interest in cardiac amyloidosis therapy and atrial fibrillation and is senior editor of, and contributor to, the recently published second edition of the book *Atrial Fibrillation: Mechanisms and Management*. He also is a fellow of the American College of Cardiology.

Faraday, Michael. (*September 22, 1791, Newington Butts, near London, UK; †August 25, 1867, Hampton Court Green, near London). English physicist and chemist. Dr. Farady made one of his greatest discoveries, that of induction, on August 19, 1831. The report on this topic initiated his series of "Experimental researches in electricity," which covered all areas of electricity that were known at that time In the years 1833 and 1834 Faraday established the electrochemical principles that were named after him and, in consultation with W. Whewell, introduced the electrochemical nomenclature that is still in use today. Faraday described all these events quite vividly with the help of the concept of magnetic and electrical lines of force that he developed. James Clerk Maxwell then combined this idea with the concept of an electromagnetic field and organized an electromagnetic theory of light.

Farré, Jerónimo. Dr. Farré studied medicine at the Universidad Complutense of Madrid, from which he graduated in 1970. After becoming a cardiologist at Fundación Jiménez Diaz of Madrid, Farré went to Maastricht, the Netherlands, for specialized education in arrhythmology and clinical electrophysiology. From October 1977 to January 1979 he stayed in Maastricht, where he was the first cardiac electrophysiology fellow to arrive at the laboratory of Prof. H. J. J. Wellens. Currently Farré is a Professor of Cardiology at the Universidad Autónoma of Madrid and a member of the Executive Scientific Committee of the European Society of Cardiology.

Fisch, Charles. Dr. Fisch was born in 1924. He is an Academic Cardiologist at Indiana University, and was a two-term President of the American College of Cardiology (ACC) (1975–1977). He, together with Leonhard Dreifus and Eliot Corday, encouraged the ACC's entry into the area of government relations. Fisch has held many positions in the ACC. In 1953, he founded the Krannert Institute of Cardiology and was its Director, as well as the Director of the Division of Cardiology at Indiana University School of Medicine until 1990. Fisch became a Distinguished Professor of Medicine at Indiana University in 1975, and Emeritus Professor in 1992. He is also an Honorary Doctor of Medicine at the University of Utrecht in the Netherlands. With Fisch's early interest in electrocardiography came an interest in the interpretation of complex cardiac arrhythmias. His brilliant bibliography focuses mostly on cardiac arrhythmias, and his three books on digitalis, electrocardiography, and electrophysiology of arrhythmias are all classics.

Fisher, John Devens. Dr. Fisher was born in Ayer, Massachusetts, on March 23, 1943. He graduated from Yale University and received his MD degree from Wayne State University in Detroit, Michigan. Postgraduate training was done at Boston City Hospital, New York Cornell Medical Center, Montefiore Medical Center–Albert Einstein College of Medicine, and the Royal Postgraduate Medical School (now Imperial University), Hammersmith Hospital in London, UK. The renowned Montefiore Pacemaker Service became the Pacer–Arrhythmia Service when Fisher joined the faculty. After a period of cooperative evolution, the Pacer and Arrhythmia (EP) components have now reunited. Fisher has served a number of institutional roles, including Chief of the Cardiology Division. He has contributed to the research literature in a number of areas, including antitachycardia pacing, serial electropharmacological testing, early ablation of Wolff–Parkinson–White, and early transvenous defibrillation. He has been a leader in several important multicenter trials, and remains active clinically and academically. Fisher is a master lecturer and one of the leading teachers and investigators in cardiac pacing and electrophysiology.

Fleckenstein, Albrecht. (*May 3, 1917, Aschaffenburg, Germany; †April 4, 1992, St. Ulrich, near Freiburg, Germany). Dr. Fleckenstein was a Professor and Head of the Department of Physiology of the Albert–Ludwigs–University Freiburg, Germany, from 1956 to 1985. He discovered the new pharmacologic principle of calcium antagonism. Fleckenstein received many honors and awards, including *"Doctor Honoris Causa"* of the Universities of Munich (Germany), Heidelberg (Germany), Limburg (the Netherlands), La Plata (Argentina), and Basel (Switzerland), as well as the ASPET Award for Outstanding Basic Pharmacological Investigations of the American Society for Pharmacology and Experimental Therapeutics (Washington, D.C. 1987), the Award for Outstanding Cardiological Research of the International Society for Heart Research (Oxford, UK 1987/1988), the Special Achievement Award of the American Society of Hypertension (New York 1988), the Distinguished Investigator Award of the American Society of Clinical Pharmacology (Baltimore, Maryland 1989), and the Albert Einstein World Award of Sciences (Canberra, Australia 1991). Fleckenstein must be regarded as one of the most significant investigators of our time.

Fontaine, Guy. Guy Fontaine was born in Corbeil Essonnes, a suburb of Paris, France. He worked at the Hôpital Jean Rostand in Ivry, France, where he is Codirector of the University Department of Clinical Electrophysiology. For the past 30 years he has continuously expanded the frontiers of electrophysiology. In 1976, he published *The Essentials of Cardiac Pacing*, which was coauthored by his mentors and colleagues, Profs. Y. Grosgogeat and J. J. Welti. Together with his talented and thoughtful surgical colleague, Dr. G. Guiraudon, Fontaine and his colleagues were the first Europeans to perform successful surgical treatment of an accessory pathway. Fontaine and his associate, Dr. Robert Frank, perfected the technique of epicardial mapping, which permitted them to obtain the first recordings of epicardial delayed potentials in humans. His work led to the discovery of arrhythmogenic right ventricular dysplasia (ARVD), which resulted in the publication of some of the first clinical descriptions of this condition.

Forssmann, Werner. (*August 29, 1904, Berlin, Germany; †June 1, 1979). Dr. Forssmann received his degree in medicine in 1929 and joined the Eberswalde Surgical Clinic, Eberswalde, Germany. During his internship he started to administer drugs directly into the heart by means of cardiac catheterization; his superiors, however, refused permission for such a risky procedure. Forssmann then practiced the procedure on

Werner Forssmann

cadavers and secretly tried it on himself. He injected a local anesthetic, punctured a vein in his left forearm, and introduced a "well-oiled" ureteral catheter. He pushed the catheter 65 cm toward his heart, walked downstairs to the X-ray room, and, with fluoroscopy, located the catheter tip. He published this pioneering experiment on himself in 1929 and described how safe the procedure was. The report was sensational, but it also drew intense criticism from German physicians who deplored him for such a "dangerous stunt." Forssman abandoned cardiology and trained in medicine. During World War II, he served as a medical officer. Until the announcement of the 1956 Nobel Prizes, he had been practicing medicine in relative anonymity. For his pioneering effort, Forssmann, together with Cournand and Richards, won the 1956 Nobel Prize for Physiology and Medicine. Later on Forssmann became head of a surgical department in Düsseldorf.

Frank, Robert. Dr. Frank was Codirector and then Chief of the Arrhythmias Centre in the Jean Rostand Hospital in Ivry sur Seine, near Paris, from 1980 to 2002. This center has been devoted to all forms of arrhythmia investigations and therapies, from pacemaker and implantable cardioverter-defibrillator (ICD) implantation to ablation. It has been among the pioneers in ablation therapies. It is the home of Stimarec, a first line alert system of pacemaker failure. The whole department recently moved to the new Institute of Cardiology in Hôpital La Salpetrière in Paris.

Franz, Michael R. Dr. Franz was born on February 15, 1949 in Eckernförde, Germany. His current position is Professor of Medicine (Cardiology) and Clinical Pharmacology at the Georgetown University Medical Center and Director of the Arrhythmia Service, Electrophysiology Laboratory and Pacemaker Center at the Veterans Affairs Medical Center in Washington, D.C. He received his postgraduate training at the Hannover Medical School, including a fellowship in cardiology. Thereafter, Franz became an Invited Fellow in Cardiology at the Johns Hopkins Hospital in Baltimore, Maryland from 1981 to 1983, and an Assistant Professor of Medicine and Codirector of the Arrhythmia Service, Cardiology Division at Stanford University School of Medicine in California from 1985 to 1991. Franz has received numerous honors and awards and has worldwide editorial responsibilities. He is a fellow of the American College of Cardiology. He has been a prolific researcher, having published pioneering papers in the most relevant peer-reviewed journals of cardiology. He is a highly sought after speaker for his articulate and thoughtful analysis of complex subjects. Franz recently edited a monograph entitled *Monophasic Action Potentials—Bridging Cell and Bedside*.

Fuchs, Leonhart. (*January 17, 1501, Wemding, near Donauwörth, Germany; †May 10, 1566, Tübingen, Germany). Physician and botanist. In 1519, he began his study of classical languages in Ingolstadt, Germany, but, in 1521, he changed his focus to medicine. He graduated in 1524, and set up practice as a physician in Munich. In 1526, he returned to Ingolstadt temporarily as a Professor of Medicine. He was a supporter of the new anatomy of Vesalius.

Funke, Hermann. Dr. Funke began implanting pacemakers in 1970 at the University Hospital, Bonn, Germany. With a hobbyist's knowledge of electronics, he followed the simplicity of asynchronous (VOO) as well as the greater complexity of demand (VVI), atrial synchronous (VAT), and atrioventricular sequential (DVI) pacemakers. He then conceived a pacing mode that combined all of these functions, atrial pacing and sensing, as well as ventricular pacing and sensing (DDD). In September 1977, he implanted a DDD pacemaker that had been made to his specified design. After showing the electrocardiographs (ECGs) to physicians, he received enthusiastic support and thus began the modern era of dual chamber pacing. He has also developed pacemaker rate modulation and antitachycardia techniques, and continues a vigorous research and development effort.

Furlanello, Francesco. Dr. Furlanello was born in 1929. From 1973 to 1996 he served as Head of the Department of Cardiology, the Arrhythmological Center at Santa Chiara Hospital, Trento, Italy, and was the founding President of the Italian Working Group on Arrhythmias. He is a consultant to the Institute of Sport Science–Italian National Olympic Committee and was consulting cardiologist to the Italian National Soccer Team during the 1990 World Cup. He was an organizer and promoter of the International New Frontiers of Arrhythmias Congresses. Furlanello is Editor-in-chief of the international journal *New Trends in Arrhythmias*.

Furman, Seymour. Dr. Furman was born in New York City in 1931, graduated from Washington Square College, New York University, and, in 1955, received his MD degree from the State University of New York, College of Medicine Downstate Medical Center. He is the father of transvenous endocardial pacing, which enabled modern electrophysiology, including diagnostic and therapeutic modalities. He was the Secretary General of the Second World Symposium on Cardiac Pacing in 1967, a founder of the North American Society of Pacing and Electrophysiology (NASPE) and NASPExAM, and has been President of both NASPE (in 1980) and NASPExAM (1985–1999). He has been Editor of *Pacing and Clinical Electrophysiology: PACE* since 1978 (see historical page 184).

Seymour Furman

Fye, W. Bruce. Dr. Fye is Chair of the Cardiology Department at Marshfield Clinic and an Adjunct Professor of the History of Medicine and a Clinical Professor of Medicine at the University of Wisconsin. He is the author of *The Development of American Physiology: Scientific Medicine in the Nineteenth Century*, and he edited *William Osler's Collected Papers on the Cardiovascular System*. Fye is a fellow of and an historian for the American College of Cardiology. He is the author of the masterful historical book *American Cardiology—The History of a Speciality and its College* (1996).

Gaita, Fiorenzo. Dr. Gaita was born in Avellino, Italy, on December 12, 1951. He graduated at the University of Turin in 1976, and became a Specialist in Cardiology in 1979 at the University of Turin (mentor: Prof. Bursca Antonino). Dr. Gaita was trained in electrophysiology at the University of Turin (1976–1980) and at the Hôpital Lariboisière in Paris with Prof. Philippe Coumel (1980–1981). At the same hospital he participated in the first implantable cardioverter-defibrillator (ICD) implant in Europe. In September 1982, Dr. Gaita performed the first transcatheter ablation of the atrioventricular (AV) node in Italy, and in 1986 the first ablation of Wolff–Parkinson–White in Italy with DC shock application. Together with Michel Haïssaguerre, he described in 1991 the procedure of transcatheter ablation of the slow pathway in patients with AV nodal reentrant tachycardias (AVNRT). Dr. Gaita published the procedure of transcatheter ablation of incessant permanent junctional reciprocating tachycardias (PJRT) with radiofrequency current in 1994. Since 1996 he has been interested in transcatheter ablation and he showed the efficacy of surgical cryoablation limited to the posterior part of left atrium. In 1998, he performed the first implant of a biventricular ICD worldwide (Asti, Italy). He described in 2002 the clinical and electrophysiological characteristics of a new syndrome: "short QT syndrome." Dr. Gaita was Director of the Department of Cardiology of the Civil Hospital in Asti, Italy, and member of the Nucleus of the Working Group on Arrhythmias of the European Society of Cardiology.

Galen. (*AD 129, Pergamon; †AD 199/200/216, Rome or Pergamon). Famous physician and, next to Hippocrates, the most important teacher of so-called classical (i.e., Greek) medicine. First, Galen studied Greek philosophy, particularly that of Aristotle. Then, in AD 146, inspired by a dream, he started to study medicine and eventually became a gladiator physician in his hometown of Pergamon. In AD 163, he came to Rome, where the Roman Caesar Marcus Aurelius had appointed him as his personal physician. This is where he gained his greatest fame. He

Galen

Galien natif de Pergame ville d'Asie, excellent Medecin
viuoit du temps des Empereurs Antonin le Philosophe
et de Commodus, on tient qu'il a vescu 140 ans.

composed an extensive compendium of all medicine known at that time, a work that contained more than 300 writings, entitled *Corpus Galenicum*.

Gallagher, John J. Dr. Gallagher was born in Brooklyn, New York, on March 3, 1943. He was strongly influenced by the pioneering work taking place in Dr. Anthony Damato's laboratory. Gallagher pioneered the electrophysiological evaluation and surgical cure of patients with Wolff–Parkinson–White syndrome and related forms of ventricular pre-excitation. At the same time, he essentially invented both the methodology for using cryoablation in arrhythmia surgery and the concept of computer-based epicardial activation sequence mapping. He also played a key role in the development and popularization of transcatheter ablation employing high-energy DC shock (see historical page 201).

Galvani, Luigi. (*September 9, 1737, Bologna, Italy; †December 14, 1798, Bologna). Physician and natural scientist. Professor of Anatomy and Gynecology in Bologna. On November 6, 1789, Galvani discovered phenomena in an experiment involving frog legs that he traced back to electrical discharges in the animal body similar to those of the Leiden

Luigi Galvani

bottle. Through this error, he directed this observation to a new field of electricity.

Gerbezius, Marcus. (*October 24, 1658, in what is now known as Slovenia; †1718). Upon completing his study of philosophy at the University of Laibach (now Ljubljana), Gerbezius studied medicine at the universities in Vienna, Padua, and Bologna, and graduated from Bologna in 1684. In 1717 he described the symptoms of bradycardia induced by complete atrioventricular (AV) block; however, these observations were not published until 1718 (posthumously). Gerbezius' descriptions preceded those of Giovanni Morgagni by 44 years. In fact, Morgagni mentioned Gerbezius several times in his work *De Sedibus et Causis Morborum per Anatomen Indagatis* when referring to the characteristics of the pulse, symptoms, and course of the disease in a patient with AV block (see historical page 137).

Marcus Gerbezius

Gillette, Paul. Dr. Gillette, a pediatric cardiologist, is a pioneer in the implantation of cardiac pacemakers and implantable cardioverter-defibrillators (ICDs), the electrophysiological study of children with congenital heart disease, and the management of childhood arrhythmias. He is a prolific author and has directed pediatric electrophysiology and device clinics in Charleston, South Carolina, and Fort Worth, Texas.

Goldschlager, Nora. Dr. Goldschlager was born and raised in New York City. She received her undergraduate degree at Barnard College, Columbia University, New York, and obtained her medical degree at New York University. Goldschlager distinguished herself with several seminal research papers on exercise stress testing, and more recently has developed an international reputation for her expertise in the areas of cardiac arrhythmias and pacemakers. Goldschlager is the recipient of the 1998 Distinguished Teacher Award of the North American Society of Pacing and Electrophysiology.

Greatbatch, Wilson. Greatbatch, an electrical engineer, designed and built the first completely implantable pulse generator in the USA in collaboration with surgeon William Chardack. Successfully implanted in

1960, this generator was the ancestor of a generation of pacers powered by mercury batteries. In the 1970s Greatbatch introduced the lithium–iodide battery, which greatly extended pacemaker longevity.

Griffin, Jerry C. Former Professor of Medicine at the University of California San Francisco and Vice President of the former Incontrol, Dr. Griffin has been President of the North American Society of Pacing and Electrophysiology (NASPE) (1985–86) and was Secretary General of the World Symposium of Cardiac Pacing and Electrophysiology (1991), held under the auspices of NASPE in Washington, D.C. He has made major contributions to epidemiology and device management of ventricular and, especially, atrial arrhythmias.

Groedel, Franz Maximilian. (*May 23, 1881, Bad Nauheim, Germany; †October 12, 1951, New York). Dr. Groedel was a pioneer of electrocardiography, cardiac radiology, and scientific hydrotherapy, and—most importantly—he was the founder of the American College of Cardiology. In 1904, he received his medical degree from the University of Leipzig (Germany). Groedel cofounded the German Society for Heart and Circulation Research in 1924. One of his main interests was clinical electrocardiography. He developed the concept of the unipolar chest lead or precordial electrode in the early 1930s, independent of Frank Wilson's group at the University of Michigan. Groedel summarized two decades of electrocardiographic research in a 1934 book that included his controversial theory that each cardiac ventricle generated an independent or "partial" electrocardiogram (ECG). His later work studies concerned the direct recording of ECGs from the surface of the heart during surgery in humans, particularly from the surface of the atria and ventricles. By 1932, he had published nearly 300 scientific articles, was a

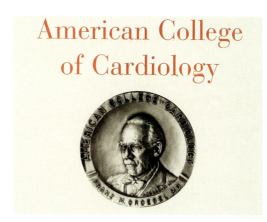

American College
of Cardiology

Franz Maximilian Groedel

Full Professor at the University of Frankfurt, Germany, had a success-ful practice in Bad Nauheim, Germany, and was Director of a world class cardiovascular research institute there. In 1933, as Hitler came into power, Groedel was classified as "non-Aryan," and he knew that his career and life were at risk; he immigrated to the USA in the same year (see historical page 162).

Guiraudon, Gerard M. Dr. Guiraudon was born and raised in Paris, France. He received his undergraduate degree and his medical training there, at the University of Paris, after which he spent 2 years as a medical officer in military service. He trained extensively in general and thoracic surgery. His contributions include innovative surgery for the ablation of ventricular tachyarrhythmias, development of an epicardial approach for interruption of accessory pathways in patients with reentrant atrio-ventricular tachycardias, cryoablation of atrial flutter, surgical exclusion of arrhythmogenic foci in the right atrium, and, of course, the "corridor" procedure for atrial fibrillation with its more recent spiral modification.

Guize, Louis J. Dr. Guize was born in 1939 in Magny-en-Vexin, France. He received his MD diploma *cum laude* from the University of Paris in 1968 with a specialization in cardiology. He was trained in the Depart-ment of Cardiology by Prof. Jean Lenègre. Guize became an Associate Professor in 1972, and a Professor of Cardiology at the University of Paris in 1981. Since 1991 his position is Head of the Department of Cardiology, firstly at Hôpital Broussais, thereafter at the Hôpital George Pompidou in Paris. Involved in experimental and clinical electrophysio-logy, particularly in sino-atrial function, antitachycardia, and hemody-namic ventricular stimulation and ablation, he received the Medtronic Award of the French Society of Cardiology in 1990. Guize is currently President of the French Society of Cardiology's Working Group on Epidemiology and Prevention, and Chairman of the IPC (*Investigations Préventives et Cliniques*) Medical Center, which manages a large cohort of men and women for cross-sectional and prospective studies, especially on cardiovascular risk factors and arrhythmias.

Gulizia, Michele Massimo. Dr. Gulizia was born on May 28, 1960 in Catania, Italy. He obtained a diploma as specialist in cardiology from the Catania University on November 25, 1988 *cum laude* and a diploma in sport medicine from the same university on October 23, 1992. In 2001, he became Medical Director of Cardiology on the first level of coronary care unit at the G. Garibaldi High-Specialization National Hospital of Catania, including the assignment for Heart Failure Diagnosis and Therapy Project. Gulizia is a member of the editorial committee of the *Giornale Italiano di Aritmologia e Cardiostimolazione*. He is mainly involved in the field of pacing and arrhythmology, and heart failure, and has published more than 175 scientific papers. He has presented more than 300 abstracts at national and international meetings and has served as invited speaker or chairman.

Haines, David Emens. Dr. Haines was born in Deerfield, Illinois, on September 1, 1954. He was educated at Williams College (BA), in Williamstown, Massachusetts (1972–1976) and the University of Rochester School of Medicine and Dentistry (MD) in Rochester, New York (1976–1980). He received his postdoctoral training at the Medical Center Hospital of Vermont, in Burlington, Vermont (medicine) and at the University of Virginia Hospital, Charlottesville, Virginia (cardiology, electrophysiology). His present position is Professor of Medicine in the Cardiovascular Division of the University of Virginia School of Medicine. He is also the Director of the Cardiovascular Fellowship Training Program in the Department of Medicine, and the Codirector of the Cardiac Electrophysiology Laboratory in the Division of Cardiology, both at the University of Virginia. He is President of the American College of Cardiology, Virginia Chapter (1999–2003) and Governor of Virginia, American College of Cardiology (1999–2003).

Haïssaguerre, Michel. Dr. Haïssaguerre was born in Bayonne, France, on October 5, 1955. He became a Professor of Cardiology in September 1994. His present position is Professor at the Hôpital Cardiologique du Haut–Lévèque, Bordeaux–Pessac. Haïssaguerre serves on the editorial boards of many major journals of cardiology, including *European Heart Journal, Circulation, Europace, The Journal of Cardiovascular Electrophysiology, Journal of Interventional Cardiology,* and *Pacing and Clinical Electrophysiology: PACE.* He has received numerous honors and awards, including the Prix Robert Debré (1982), the Prix de l'Information Cardiologique (1990), and the Prix Ela Medical (1992). His scientific and clinical work focuses on cardiovascular electrophysiology, particularly ablation of atrial fibrillation. He is best known for his remarkable contributions in the area of atrial fibrillation ablation. He was the first to detect the importance of pulmonary vein triggers and drivers in the genesis of atrial fibrillation. In addition, he was first to propose the technique of pulmonary vein isolation, which underlies current methods used throughout the world for atrial fibrillation cure. He and his colleagues have extended these observations to include ablative lesions between the veins and to the mitral annulus. Haïssaguerre has published more than 240 publications in the leading peer-reviewed cardiology journals dealing mainly with radiofrequency current endocardial ablation of tachyarrhythmias. In 2004, Haïssaguerre received the Pioneer in Cardiac Pacing and Electrophysiology award by the North American Society of Pacing and Electrophysiology (NASPE)–Heart Rhythm Society.

Hammill, Stephen C. Dr. Hammill was born in Denver, Colorado, on February 26, 1948. His present positions are Professor of Medicine at the Mayo Medical School and Director of the Electrocardiography Laboratory, Director of the Electrophysiology Laboratory, and Consultant in the Division of Cardiovascular Diseases at the Mayo Clinic in Rochester, Minnesota. His many professional positions and appointments include President of the Minnesota Chapter of the American College of Cardiology (1991–1994), member of the editorial board of the *British Heart Journal* (1994–1999), second Vice President of the North American Society of Pacing and Electrophysiology (NASPE) (2001–2002), Chair of the Health Policy Committee of NASPE, and Board of Trustees of NASPE. Hammill's scientific publications and most recent presentations concern arrhythmology, particularly diagnostic procedures, radiofrequency therapy, catheter ablation of supraventricular tachycardias, antiarrhythmic drug treatment of syncope, and risk assessment in athletes.

Harken, Dwight. Dr. Harken was of the founding generation of cardiac surgeons. During World War II, he developed a technique of closed cardiac surgery to successfully remove intracardiac and intrapericardial foreign bodies such as bullets and shrapnel. During the postwar period he was a developer of closed mitral commissurotomy, the first commonly performed repair of acquired cardiac lesions. He implanted a "demand" pacemaker early after its design and initial construction.

Harthorne, J. Warren. Dr. Harthorne graduated from Bowdoin College in Maine, and went on to McGill University in Canada to obtain his medical degree. After a medical residency at Montreal General Hospital and a tour of duty as a medical officer in Korea, Harthorne undertook his cardiology training at Massachusetts General Hospital. He remains there as an Associate Professor of Medicine at Harvard Medical School and a physician at Massachusetts General Hospital. In the 1970s he played a leading role in the founding of the North American Society of Pacing and Electrophysiology (NASPE) and served as the organization's first and longest-acting President. Harthorne is the recipient of NASPE's Founders Award, 2001. This is his second award from NASPE in recognition of his pivotal role in establishing what is regarded today as the preeminent organization in the world dedicated to the study and management of arrhythmias.

Harvey, William. (*April 1, 1578, Folkestone, Kent, UK; †June 3, 1657, Hampstead, London). Physician and anatomist. Harvey's pioneering achievement was the experimental discovery of the (greater) circulation

William Harvey

of blood, which he succinctly described in his work *Exercitatio Anatomica de Motu Cordis et Sanguinis in Animalibus* (Frankfurt a.M., 1628). His embryological studies represent another great achievement, which he recorded in his *Exercitationes de Generatione Animalium* (London, 1651).

Hatala, Robert. Dr. Hatala is currently Director of the Department of Arrhythmias at the Slovak Cardiovascular Institute in Bratislava, which he founded in 1995 as the first specialized high-volume arrhythmia center in the Slovak republic. He is currently professor of internal medicine—cardiology at the Comenius University School of Medicine in Bratislava, Slovakia. Hatala obtained his MD degree (*summa cum laude*) and his PhD degree at the Comenius University School of Medicine in Bratislava in 1980 and 1988, respectively. His main interest is devoted to all aspects of clinical cardiac electrophysiology. In this field he worked as fellow and later on as visiting professor at several institutions in Europe and North America (CHU Pitié–Salpêtrière, Paris, France; University Hospital Hamburg—Eppendorf, Germany; Ludwig Boltzmann Institute, Vienna, Austria; Hospital of Sacré-Coeur, University of Montreal, Canada). He has authored more than 200 presentations in medical journals and on major international cardiological meetings. As an invited speaker he has lectured in several languages including English, French, and German. In 1996 and in 2000, he was elected President of the Slovak Society of Cardiology.

Hauer, Richard N.W. Hauer is the Director of the Clinical Electrophysiology Laboratory and the Cardiac Arrhythmia Unit at the Heart–Lung Institute of the University Hospital, Utrecht, the Netherlands, where he was also appointed Full Professor of Clinical Electrophysiology in 1996. After receiving his MD degree at Leiden State University in 1974, he completed a fellowship in general internal medicine in the Hague. In 1980, he was registered as a cardiologist and worked for the next 2 years at a private hospital in Utrecht. In 1982, he left the Netherlands for the USA, where he received further training during a fellowship in clinical electrophysiology at the Krannert Institute of Cardiology in Indianapolis, Indiana. His thesis, submitted in 1987, was entitled "The site of origin of ventricular tachycardia: Identification, localization and ablation using catheter techniques."

Hauser, Robert G. Former President of North American Society of Pacing and Electrophysiology (NASPE), CEO of a major device manufacturer, and member of the NASPExAM writing committee, Dr. Hauser is a practicing cardiologist and presently operates a pacemaker and implantable cardioverter-defibrillator registry via Internet communication.

Hayes, David L. Dr. Hayes was born on September 20, 1953 in Rolla, Missouri. His present titles are Consultant in Cardiovascular Diseases and Internal Medicine, Vice Chair of the Cardiovascular Division, and Director of Pacemaker Services at the Mayo Clinic and Professor of Medicine at the Mayo Medical School in Rochester, Minnesota. Hayes graduated from the University of Missouri, Kansas City School of Medicine in December 1976. He undertook a year of special cardiology training at New York Hospital—Cornell doing nuclear cardiology with Dr. J. Borer, and 6 months at Montefiore Medical Center with

David L. Hayes

Dr. S. Furman. He received the North American Society of Pacing and Electrophysiology traveling fellowship in 1983, and spent 3 months at Clinic Val d'or with Dr. Jacques Mugica. Hayed is the first recipient of the E. Grey Dimond "Take Wing" Award and a past President of the North American Society of Pacing and Electrophysiology (1998–1999).

Herophilus. (*ca. 330 BC, Chalcedon; †ca. 250 BC). Physician and anatomist. Herophilus worked in Alexandria, the early Hellenic cultural metropolis. He was a supporter of the theory of four humors and emphasized the importance of "empeiria" (experience) for medicine. His extraordinary anatomical research on the human body represents a milestone in the development of anatomy before the Renaissance. He described individual organs and body structures for the first time as well as the difference between arteries and veins.

Hippocrates. (*ca 460 BC, Island of Cos; †ca. 370 BC, Larissa). Physician, legendary founder of Greek medicine. As the spiritual rector of the Hippocratic manuscript collection (*Corpus hippocraticum*), he gained paramount importance in the medical history of the West for establishing the medical profession's tradition of education. As an "Asclepiad" and head of the School of Cos, he was thought to have been, according to tradition, a nineteenth generation descendent of Asclepius (see historical page 135).

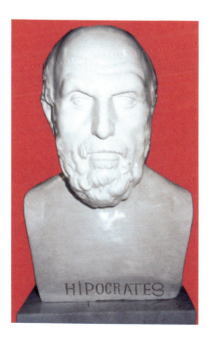

His, Wilhelm, Jr. (*December 19, 1863, Basel, Switzerland; †November 10, 1934, Brombach, Germany). His was the son of Wilhelm His (*1831; †1904), a Professor of Anatomy and Physiology in Leipzig, Germany, and one of the most respected physiologists and anatomists of his time. After completing studies in Leipzig, Bern, and Strasbourg, Wilhelm His, Jr. received his doctorate in medicine in Leipzig in 1889. He served his internship at the Medical Hospital of Leipzig, where he was awarded his degree in Internal Medicine in 1891 and appointed Associate Professor in 1895. From 1902 to 1907 he held professorships at Basel, Göttingen, and Berlin. His extensive publications on numerous subjects in the field of internal medicine focus primarily on heart disease and metabolic disorders. In his work "Embryonic cardiac activity and its significance for adult heart movement theory," published in 1883, he described the anatomical feature named after him, the bundle of His. He also worked in the field of medical history, publishing the *History of the Medical Hospital in Leipzig* in 1899.

Hoffman, Brian. A pioneer in microelectrode recording of cardiac electrograms, Dr. Hoffman, in 1960, published the classic *Electrophysiology of the Heart* with Paul Cranefield. He then described the slow component of conduction in reentry pathways and slow response action potentials in living tissue that permit reentry. He was Chair of Pharmacology at Columbia University between 1963 and 1996, and trained two generations of electrophysiologists who are now themselves chairs of departments throughout the world.

Hoffmann, Ellen. Dr. Hoffmann was born on November 17, 1958 in Leverkusen, Germany. She currently is an Associate Professor at the Ludwig–Maximilian University in Munich, Germany. She received her education at the University of Heidelberg in Germany, and at the Universities of Oxford and Cambridge in the UK. She holds the position of Senior Consultant Cardiologist, and received the 1998 Excellence in Teaching Award from the Bavarian Government as well as the Therese von Bayern Foundation Prize for scientific achievement in 2000. Hoffmann's primary area of interest is electrophysiology, with a focus on catheter ablation and implantable cardioverter-defibrillator (ICD) technologies, drug treatment and pacing for prevention of atrial fibrillation, and arrhythmogenomics.

Hohnloser, Stefan. Dr. Hohnloser, born June 2, 1954 in Pforzheim, Germany, received his medical training at the University of Freiburg. From 1980 to 1994 he was a research assistant at the University of Freiburg, initially in the Department of Physiology and subsequently in the Department of Internal Medicine. He then spent 2 years at the Harvard School of Public Health and Brigham Women's Hospital, Boston, Massachusetts, before returning to Freiburg. In 1995, he moved to Frankfurt, where he is currently Professor of Medicine and Cardiology at the J. W. Goethe University and Director of the Electrophysiology Laboratories and Clinical Arrhythmia Service. Hohnloser's research interests include sudden cardiac death, risk stratification after acute myocardial infarction, the use of device therapy for life-threatening arrhythmias, and the pharmacological and non-pharmacological treatment of atrial fibrillation.

Hollman, Arthur. Dr. Hollman was born in 1923. He is a British cardiologist who studied under Thomas Lewis at University College Hospital in 1943. He was among the first British cardiologists to perform cardiac catheterization (beginning in 1957 at Hammersmith Hospital, London, UK).

Holzmann, Max. (*March 31, 1899; †January 27, 1994). Dr. Holzmann studied medicine at the Universities of Zürich and Lausanne in Switzerland. After his final medical examination and approval of his thesis (Doctor of Medicine) in 1923, he began his professional training in Vienna, Austria, then in Paris, France, followed by a 6-year internship at the University Clinic of Zürich (with Prof. Otto Naegeli). Holzmann worked in his private cardiology practice in Zürich from 1932 on, and he received the title of Habilitation/Assistant Professor in 1960. His special interests and research subjects were radiology of the circulatory system and electrocardiology. Between 1942 and 1965 he authored numerous original manuscripts and reviews as well as five editions of his famous textbook *Klinische Elektrokardiographie* (Clinical Electrocardiography).

Hori, Motokazu. Dr. Hori was trained as a cardiac surgeon and became involved with cardiac pacing in 1960, upon the opening of a unit, which he helped to build, at the University of Tokyo. He presented at the Fourth World Symposium in Groningen in 1973, and was a founder of the International Cardiac Pacing Society and then the Secretary General of the Fifth World Symposium in Tokyo in 1976.

Ideker, Raymond E. Dr. Ideker was born on July 31, 1942 in Oak Park, Illinois. His present titles are Jeanne V. Marks Professor of Medicine, Professor of Biomedical Engineering, and Professor of Physiology. Ideker received his education at the University of Tennessee in Memphis and Duke University Medical Center in Durham, North Carolina. He is the coinventor of 21 US patent-holding medical devices. With 218 peer-reviewed publications and more than 50 book chapters to his credit, Ideker is also a highly sought-after speaker because of his articulate and thoughtful analysis of complex studies. Undoubtedly, the medical field will continue to profit greatly from his prolific research and generous mentoring.

Irnich, Werner. Dr. Irnich, a Professor of Biomedical Engineering, has analyzed chronaxie and rheobase and electrode function. His work has supported our modern understanding of the basic laws of cardiac stimulation. He has described the concept of dual-chamber universal pacing, has analyzed the effect of interference signals generated by cellular telephones and other devices on cardiac pacemakers and defibrillators, and has conducted a registry for pacemakers removed from patients to correlate pacemaker function and possible cause of death.

Iwa, Takashi. A cardiac surgeon, Dr. Iwa is a pioneer in the field of cardiac arrhythmia surgery. He first implanted pacemakers from 1960 to 1962, and used external DC countershocks for the management of atrial fibrillation shortly thereafter. Iwa designed an atrial radiofrequency pacemaker for supraventricular tachycardias, conceived of simultaneous implantation of both a pacemaker and a defibrillator for ventricular tachycardia followed by fibrillation, mapped the endocardium, and pioneered operations for the cure of Wolff–Parkinson–White syndrome.

Jackman, Warren M. Dr. Jackman was born in Miami, Florida, on March 11, 1952. He attended the Georgia Institute of Technology and the University of Florida, from which he graduated with an MD degree in 1976. Catheter ablation was in its infancy when he reported in 1983 the first attempt to ablate an accessory pathway with high-energy shocks. He was guided by an electrogram, and achieved temporary elimination of accessory pathway conduction. When radiofrequency current was introduced to replace the riskier shocks, the field moved slowly because of the small lesions with standard catheters. Jackman made the small step that was actually a giant leap: the large-tip electrode. He then presented the first large series of cure of atrioventricular reentrant tachycardia by radiofrequency catheter ablation. This curative treatment, based largely on Jackman's methods, is now practiced routinely worldwide.

Jaïs, Pierre. Dr. Jaïs was born on February 16, 1964. He received the degree of Doctor of Medicine at the University of Bordeaux, in

France, in 1993. In 1994, he became Chef de Clinique des Universités—Université de Bordeaux. Jaïs received the Prix Medtronic in 1997 and the Prix Nativelle in 2001. Presently he is actively involved in cardiac electrophysiology at the University of Bordeaux, Hôpital Haut Lévêque, Bordeaux–Pessac, France. Jaïs has published 21 scientific articles in major peer-reviewed journals, book chapters on electrophysiology and radiofrequency catheter ablation, and four papers in *Circulation* as the primary author.

Jalife, José. Dr. Jalife was born in Mexico City, Mexico, on March 7, 1947. His present position is Professor and Chairman of Pharmacology and Professor of Medicine and Pediatrics at SUNY Upstate Medical University in Syracuse, New York. Jalife is a master lecturer and one of the leading figures in the cardiac electrophysiology community. He received his education and training at the Escuela Nacional Preparatoria, Mexico, School of Medicine, National University, Mexico, and at the General Hospital, Oviedo, Spain. Jalife has received numerous honors and awards as well as honorary memberships. He is "Doctor Honoris Causa" at the University of Buenos Aires, Argentina. His awards include the Young Investigator Award from the American College of Cardiology (1979), the Research Award from the American Heart Association (1991), and the President's Award for Excellence and Leadership in Research from the SUNY Health Science Center at Syracuse, New York (1998). He recently received an award from the Académie Royale de Médecine de Belgique and was honored with the Distinguished Scientist Award from the American College of Cardiology (2001).

Janse, Michiel J. Dr. Janse was born in Amsterdam in 1938. His interest in cardiac electrophysiology began in the laboratory of Professor Durrer in Amsterdam in 1959, when he was 21 years old, and was further nurtured during early training in New York City in 1962 with Brian Hoffman. A year later, he returned to his birthplace, finished his medical training, and embarked on a long collaborative research relationship with Durrer until the latter's death in 1984. In 1975, Janse became Chief of the Laboratory of Experimental Cardiology, and in 1985 was appointed Professor of Experimental Cardiology at the University of Amsterdam. Janse has explored the electrophysiology of the atrioventricular (AV) node many times during his theses of 25 years and, in 1969, demonstrated that the direction of the cardiac impulse into the AV node altered AV nodal transmission. Several years later, he correlated AV nodal structure and function by relating morphological with electrophysiological findings in the rabbit AV node. Janse is recipient of the 1993 Distinguished Scientist Award of the North American Society of Pacing and Electrophysiology. Currently, he is Editor-in-chief of *Cardiovascular Research*.

Jordaens, Luc J.L.M. Dr. Jordaens was born in Essen, Belgium, on January 10, 1954. He studied in Antwerp and Ghent and became a medical doctor in 1979. He has been a specialist in cardiology since 1984. In Ghent, Dr. Jordaens obtained a Biomedical Doctorate (1987) and an Aggregation thesis in 1990: "Studies in patients with life-threatening ventricular tachyarrhythmias." After teaching in Ghent, he became Associate Professor at the Erasmus Medical Center in Rotterdam, the Netherlands. He worked on low-energy shock ablation and introduced implantable cardioverter-defibrillator (ICD) therapy and radiofrequency ablation in Belgium. The first ICD implantation in a catheter laboratory in Europe was performed under his direction. He published on pacing, invasive and non-invasive electrophysiology, sudden death, atrial fibrillation, and imaging in these areas. He was Chairman of the Belgian Society of Cardiology and Nucleus Member of the European Working Group on Arrhythmias. Dr. Jordaens is Coeditor of *Europace* since its foundation.

Josephson, Mark E. Dr. Josephson graduated from Trinity College and Columbia University College of Physicians and Surgeons in New York in 1969. His internship and residency were served from 1969 to 1971 at Mount Sinai Hospital in New York. From 1971 to 1973 he was a Research Associate at the US Public Health Service Hospital in Staten Island. He completed his cardiology fellowship at the University of Pennsylvania School of Medicine and joined the faculty as Director of Clinical Electrophysiology. Josephson was a faculty member of the University of Pennsylvania School of Medicine from 1975 to 1992. From 1981 to 1991 his academic appointment was Robinette Professor of Medicine. In 1992, he became Director of the Harvard Thorndike Electrophysiology Institute and Arrhythmia Service of Beth Israel Hospital, Boston, and Professor of Medicine at Harvard Medical School. During the past 20 years, Josephson has been a major contributor to the growing body of knowledge pertaining to the mechanisms, pathophysiology, and treatment of ventricular tachyarrhythmias. He standardized left ventricular mapping, which eventually made possible catheter ablation of ventricular tachycardia. He also introduced and directed subendocardial resection for the surgery of ventricular tachycardia. His work serves as a solid foundation for today's sophisticated therapy of ventricular arrhythmias. Josephson is a master teacher of electrocardiography. In 2001, he was honored as a Pioneer in Pacing and Electrophysiology by the North American Society of Pacing and Electrophysiology in Boston.

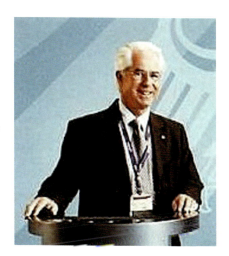

Lukas Kappenberger

Kappenberger, Lukas. Dr. Kappenberger, born on October 4, 1943, is Medical Director of Cardiology at Centre Hospitalier Universitaire Vaudois (CHUV) in Lausanne, Switzerland. He received his specialized medical education at the Medical Clinic, University Hospital of Zürich. With Edgar Sowton, he published works on sick sinus syndrome, long-term electrocardiograms (ECGs), and programmed pacing. In 1982, he received the highest award of the Swiss Society for Internal Medicine. His special areas of interest include clinical electrophysiology in diagnostics and therapy and differentiated pacemaker therapy. Kappenberger has lead many important studies, including one on obstructive cardiomyopathy and its treatment with electrostimulation. He is extensively involved in the scientific planning of the European Society of Cardiology.

Kautzner, Josef. Dr. Kautzner was born on November 9, 1957, in Vlasim, Czechoslovakia (now the Czech Republic). He graduated at Charles University Medical School I. in Prague (1983), and trained in the Charles University General Hospital, Prague, and in St. George's Hospital, London, UK. He is specialist in cardiology/electrophysiology, and currently Head of the Department of Cardiology at the Institute for Clinical and Experimental Medicine in Prague, and Associate Professor at Charles University Medical School I. in Prague. Kautzner is internationally recognized as a Fellow of the European Society of Cardiology and regular member of the North American Society of Pacing and Electrophysiology (NASPE)–Heart Rhythm Society. He is the principal investigator for several grant projects. Kautzner is actively involved in catheter ablation, cardiac resynchronization and implantable cardioverter-defibrillator (ICD) programs. His publication activity covers 64 peer-reviewed papers, 40 reviews, 20 book chapters, and one monograph.

Kenda, Miran F. Dr. Kenda was born in Celje, Yugoslavia (now Slovenia), on March 8, 1936. His hospital address is University Medical Center, Clinic of Cardiology, Zaloška 7, SI-1000 Ljubljana, Slovenia. Kenda received his medical education at the Medical Faculty, University of Ljubljana (1954–1960), and his special training in internal medicine–cardiology at the Clinic of Internal Diseases, Ljubljana (1965–1969). His doctor's thesis (PhD) was completed at the Medical Faculty, University of Ljubljana in 1976. Kenda was appointed as Chief of the Coronary Department, Clinic of Cardiovascular Diseases (1969), and in 1979 as Chief of Cardiology, Department A., University Medical Center, Ljubljana. In 1996, he became senior councilor. In 1985, he was appointed as full professor in internal medicine in the field of cardiology, Medical Faculty, University of Ljubljana. Kenda received numerous scientific merits and awards including honorary membership of the Hungarian Society of Cardiology (1995), honorary membership of the Croatian Society of Cardiology (1998), and European cardiologist diploma (2000). Kenda served in four editorial boards of renowned scientific journals. He is author or coauthor of more than 200 scientific articles and editor of several proceedings of the scientific meetings of the Slovenian Society of Cardiology.

Kisch, Bruno. (*August 28, 1890, Prague, in what is now known as the Czech Republic; †August 12, 1966, Bad Nauheim, Germany, buried in Jerusalem). Prof. Kisch was Head of the Department of Physiology, Biochemistry, and Pathologic Physiology at the University of Cologne (1925). He founded the German Society of Cardiology on June 3, 1927, in Bad Nauheim. Later, he was forced to leave his academic position because he was considered a "non-Aryan" by the Nazi Government. He went into medical practice, treating cardiovascular diseases, and immigrated to the USA in 1938. His special interests included cardiac arrhythmias and potassium metabolism. He was "a man of unlimited curiosity" and the inventor of the electrophysiological phenomenon called "overdrive suppression." Kisch published many pioneering articles and books on the disorders of cardiac rhythm. He coauthored numerous papers with Franz M. Groedel—also an immigrant from Germany and the founder of the American College of Cardiology (ACC). Kisch served as the ACC President from 1951 to 1953.

Kléber, André G. Dr. Kléber was born in Neuchâtel, Switzerland, on December 9, 1944. His laboratory address is Department of Physiology,

André Kléber

University of Berne, CH-3012 Bern, Switzerland. He received his basic training at the Department of Physiology, University of Berne (Prof. Silvio Weidmann) and in the Department of Clinical Physiology, University of Amsterdam, the Netherlands (Prof. Dirk Durrer). In 1996, he became Professor of Physiology, University of Berne. Kléber has published more than 75 papers in the leading peer-reviewed experimental and clinical journals of cardiology and served on the editorial boards of the *Journal of Molecular and Cellular Cardiology*, *The Journal of Cardiovascular Electrophysiology*, *Cardiovascular Research*, and *Circulation Research*. Kléber received relevant honors and awards including the Prize of the Swiss Heart Foundation (1985), the Albrecht von Haller Medal, University of Berne (1986), and the Dirk Durrer visiting professorship award, University of Amsterdam (1996). Kléber is really polyglot and speaks German, English, French, Dutch, Italian, and Russian. He is married to Barbara Ann (Kléber) and has three sons Maurice, Philippe, and Marc-André.

Klein, George Joseph. Dr. Klein was born in Budapest, Hungary, on June 7, 1947. His present position is Chairman, Division of Cardiology at the University of Western Ontario and Faculty of Medicine, Department of Medicine at London Health Science Centre, University Campus, in London, Ontario, Canada. Klein is an active member of the editorial boards of several journals, including *The American Journal of Cardiology*,

Canadian Journal of Cardiology, Circulation, Journal of the American College of Cardiology, American Heart Journal, and *European Heart Journal.* He is Associate Editor of *Pacing and Clinical Electrophysiology: PACE.* He served on the Board of Trustees of the North American Society of Pacing and Electrophysiology from 1991 to 1995. Since 2000 Klein has been Chairman of the Arrhythmia Section of the American Board of Internal Medicine. Klein has published five books, 88 book chapters, and 336 scientific articles on cardiology, particularly arrhythmology. He holds eight patents.

Klein, Helmut. Dr. Klein was born in 1941 in Berlin. He studied medicine in Göttingen, Düsseldorf, and Bonn. From 1978 to 1980 he was a Fellow of the Max Kade Foundation in New York, as well as a Fellow at the University of Alabama in Birmingham (Prof. A. Waldo, Prof. J. Kirklin, Prof. T. James). His thesis was devoted to intraoperative mapping, catheter mapping, and programmed pacing for ventricular tachycardias. Since 1992 he has been the Chief of the Hospital for Cardiology, Angiology, and Pneumology at the Centre for Internal Medicine at the Otto von Guericke University, Magdeburg. His scientific work involves clinical and interventional electrophysiology, the development of defibrillator therapy, cardiomyopathy (morphology and metabolism), and cardiological intensive care medicine. In 2004, Klein received the Michel Mirowski Award (sponsored by Guidant).

Kowey, Peter R. Dr. Kowey received his degree in medicine in 1975 from the University of Pennsylvania. His clinical and professional experience includes a fellowship from 1978 to 1980 at the Harvard University School of Public Health and Peter Bent Brigham Hospital, Cardiology. Kowey is now a Professor of Medicine at Jefferson Medical College and Chief of the Division of Cardiovascular Diseases at the Lankenau Hospital and Medical Research Center in Pennsylvania. He also holds professorships at the Medical College of Pennsylvania and Allegheny University of Health Sciences. His principal area of interest has been cardiac rhythm disturbances. He is a founding member of the Philadelphia Arrhythmia Group and a charter member of the North American Society of Pacing and Electrophysiology. He spent 9 years as a member of the Cardiorenal Drug Advisory Committee and 4 years on the Cardiovascular Devices Committee of the Food and Drug Administration.

Krikler, Dennis, M. Dr. Krikler was born in 1928. He is a renowned British academic cardiologist at Hammersmith Hospital in London, UK, and a former Editor of the *British Heart Journal.* He has been active in the

Dennis M. Krikler

American College of Cardiology and the American Heart Association, and has provided a European perspective on American cardiology.

Kuck, Karl-Heinz. Dr. Kuck was born on April 20, 1952 in Aachen, Germany. From 1980 to 1981 he was a member of the Department of Clinical Electrophysiology at Limburg University Hospital in Maastricht, the Netherlands (with Prof. H. J. J. Wellens). His thesis pertained to electrophysiology of hypertrophic cardiomyopathy. In 1989, he became an Associate Professor in the Department of Cardiology at the University Hospital in Hamburg—Eppendorf. Kuck is one of the most renowned German electrophysiologists and has a prominent international reputation. He is especially noted for his work with radiofrequency catheter ablation of accessory atrioventricular pathways. Currently, Kuck is the Head of the II. Medical Clinic of St. George General Hospital in Hamburg, Germany.

Laënnec, Théophile Réné Hyacinthe. (*February 12, 1781, Quimper, France; †August 13, 1826, Kerlouan, France). Internist. Known as the founder of auscultation, he studied medicine in Nantes, and in 1800 attended the Ecole de Médecine in Paris, where he graduated in 1804. He introduced auscultation with the help of the stethoscope, which he invented in 1819, an instrument that revolutionized the clinical diagnostics of thoracic diseases. He, among others, was the first to recognize tuberculosis of the lungs as an autonomous disease by describing the

T.R.H. Laënnec

disease in precise macroscopic and pathological terms. He himself died of this disease at the age of 45 years.

Lagergren, Hans. Using equipment built by the Elema–Schönander Company in Stockholm, Sweden, surgeon Hans Lagergren explored the use of transvenous leads for chronic cardiac pacing in the early 1960s. He implanted the first transvenous pacemaker in 1962. His reports greatly accelerated the transition from myocardial to transvenous pacing.

La Rovere, Maria Teresa. Dr. La Rovere was born on September 8, 1952 in Orta Nova (Foggia), Italy. She graduated in Medicine at the University of Milan in 1978, and became a specialist in cardiology in 1980. She was visiting physician at the Department of Cardiovascular Medicine, John Radcliffe Hospital, Oxford, UK, at the Department of Cardiology of the Presbyterian Hospital, Columbia University, New York, and at the Laboratory of Electrophysiology, St. Georg Hospital, Hamburg, Germany. Currently she is Head of the Laboratory of Autonomic Nervous System Analysis at the Department of Cardiology of the Fondazione "Salvatore Maugeri," IRCCS, Pavia, Italy, and Teaching Professor at the School of Cardiology, University of Pavia, Italy. Her main research interests involve the pathophysiology of the autonomic nervous system in ischemic heart disease and in heart failure, the development of clinical tools for the study of the autonomic nervous system, and the effects of pharmacological and non-pharmacological interventions on autonomic balance. Her work has contributed to defining the role of autonomic dysfunction in the identification of postmyocardial infarction patients at high risk for sudden and non-sudden cardiac death. She serves the leading cardiology journals as a reviewer. Her publications include 40 chapters and more than 100 original papers in peer-reviewed journals.

Arne Larsson

Larsson, Arne. A. Larsson was born May 26, 1915 in Stockholm, Sweden. While in his mid thirties, he developed myocarditis as a result of food poisoning he contracted from eating oysters and from having hepatitis. The result was complete heart block. Consequently, Larsson was bedridden for 2 years before his wife persuaded surgeon Åke Senning to implant an experimental pacemaker that Senning and the engineer Rune Elmqvist had been studying. The year was 1958, and Larsson's rechargeable pacer failed quickly. Nevertheless, he survived and was given a more reliable model some years later. Larsson received a total of 27 pacemaker devices. He died on December 28, 2001.

Lau, Chu Pak. Dr. Lau was born on April 4, 1957 in Hong Kong. His present position is Professor and Chief of the Cardiology Division of the Department of Medicine at the University of Hong Kong. He received his medical training at the University of Hong Kong from 1976 to 1981. His postdoctoral training was done at St. George's Hospital in London, UK. Lau's scientific merits include one textbook, 20 chapters in books, and more than 200 scientific publications in major peer-reviewed journals in the field of pacing and electrophysiology. His main areas of interest are the development of rate-adapted pacing, atrial defibrillators, and cryoablation of cardiac arrhythmias. Lau received the Coucher Foundation Fellowship and, in 1997, was listed as one of the 10 Most Outstanding Persons of Hong Kong. He was Secretary General of the Eleventh

World Congress of Cardiac Pacing and Electrophysiology, 2003, in Hong Kong.

Lazzara, Ralph. Dr. Lazzara was born in Tampa, Florida, on August 14, 1934. He received his BA degree in 1955 from the University of Chicago, and his MD degree in 1959 from Tulane Medical School in New Orleans, Louisiana. After learning the microelectrode technique with Dr. Brian Hoffman at Columbia University College of Physicians and Surgeons, he returned to New Orleans to direct the Cardiovascular Research Laboratories at the Ochsner Foundation Hospital before entering the Army Medical Corps (1967–1970). In 1978, as an independent investigator, Lazzara published pioneering microelectrodes studies on the proarrhythmic actions of lidocaine. These reports predated the clinical coining of the term "proarrhythmia." In tribute to his accomplishments and continued devotion to basic and clinical inquiry, Lazzara received the 1990 Distinguished Scientist Award from the North American Society of Pacing and Electrophysiology.

Le Heuzey, Jean-Yves. Dr. Le Heuzey was born on May 12, 1951 in Neuilly, France. He was an intern at the Hôpitaux de Paris in 1974, and received his MD degree and the title of Specialist in Cardiology in 1980. In 1990, he became a Professor of Cardiology at the Pierre and Marie Curie University in Paris. Le Heuzey was Chairman of the Working Group on Cardiovascular Research of the French Society of Cardiology in 1994; he became the Chairman of the Society's Working Group on Arrhythmias in 2000. Le Heuzey became a fellow of the European Society of Cardiology in 2001, and is Head of the Laboratory of Experimental Electrophysiology, Broussais and Georges Pompidou Hospitals in Paris. He has authored approximately 130 original publications in peer-reviewed journals.

Lesh, Michael. Dr. Lesh is a world renowned cardiac electrophysiologist, researcher and founder of three medical device companies based on his inventions: Atrionic, MitraLife, and Appriva. He earned degrees from the Massachusetts Institute of Technology and the University of California, San Francisco (UCSF) School of Medicine. At UCSF, he was Associate Professor of Medicine and Chief of Cardiac Electrophysiology. Lesh was a pioneer in Ê methods catheter ablation of cardiac arrhythmias, and author of over 300 scientific papers. At Atrionix, a developer of an ultrasound ablation catheter for curing atrial fibrillation, Lesh was Director and Chief Science Officer. Atrionix was acquired by Johnson and Johnson. Lesh was Chairman and CEO of MitraLife, which is developing a catheter-based treatment of congestive heart failure, and of Appriva, which is in clinical trials of PLAATO, a percutaneous transcatheter device for preventing stroke. Both Appriva and Mitralife have been acquired by EV3. Currently, Dr. Lesh is Chairman and CEO of MVM Productions creating motion picture films. He also finds great joy in spending time with his family, especially his 6-year-old son, Jonathan, and his $2^{1}/_{2}$-year-old daughter, Rebecca.

Lévy, Samuel. Dr. Lévy is currently a Full Professor of Medicine and Cardiology at the University of Aix–Marseille School of Medicine and Chief of the Division of Cardiology at Hôpital Nord in Marseille, France. After medical and postgraduate training at the University of Paris School

Samuel Lévy

of Medicine, Lévy completed several fellowships in internal medicine and cardiology at hospitals in France, as well as a senior fellowship at the University of Miami School of Medicine. In 1975, he became an Associate Professor of Cardiology at the University of Bordeaux, and in 1979 he became a Professor of Medicine and Cardiology at the University of Marseille School of Medicine. Lévy received the university's highest promotion in 1991, to Professor First Class of Cardiology. He is a noted researcher in the study of atrial fibrillation and electrophysiology, having published numerous related scientific papers and articles. He is a past Chairman of the Working Group of Arrhythmias of the European Society of Cardiology and a fellow of the European Society of Cardiology as well as a fellow of the American College of Cardiology.

Lewalter, Thorsten. Dr. Lewalter was born on October 18, 1964 in Fulda, Germany. He performed his medical training at the Universities of Marburg and Würzburg, Germany, followed by an internship and residency at the University of Bonn, Germany, where he also got his training in cardiology and clinical cardiac electrophysiology. Lewalter became Director of the Electrophysiological Laboratories at the University Hospital Bonn in 2000, and was appointed as an Assistant Professor in 2001. He is a member of the German Cardiac Society and the Working Group on Arrhythmias of the European Society of Cardiology, and, in addition, a member of the editorial board of the *Zeitschrift für Kardiologie* and *Herzschrittmacher und Elektrophysiologie*. Lewalter has received the North American Society of Pacing and Electrophysiology (NASPE) Travel Award for Fellows in 1996 and was repeatedly selected as a faculty member of the NASPE scientific sessions. His research interests include tachycardia mechanisms, mapping techniques and catheter ablation of supraventricular and ventricular tachyarrhythmias, pharmacological and interventional therapy of atrial fibrillation, and pacemaker and implantable cardioverter-defibrillator (ICD) therapy.

Lewis, Thomas. (*1881; †1945). A pioneer in the fields of clinical electrocardiography and cardiac electrophysiology, Dr. Lewis was a Welshman, educated in Cardiff, Wales, UK, and at University College Hospital in London, UK, where he spent his entire career. Starting in 1909 he used Einthoven's newly invented string galvanometer to do an intensive study of arrhythmias, and his 1911 monograph *The Mechanism of the Heart Beat* was hailed as the 'bible' of electrocardiography. Its third, renamed edition of 1925 is one of the classics of cardiology. Willem Einthoven became his good friend and valued collaborator. Using the 1911 Cambridge Instrument Company's electrocardiograph, Lewis studied

the spread of the excitation wave in a dog heart, using even endocardial electrodes, which laid the foundation for cardiac electrophysiology. In 1909, together with Sir James Mackenzie, he founded and became the Editor of the journal *Heart*. His books *Clinical Electrocardiography* (1913) and *Diseases of the Heart* (1933) were widely acclaimed and went through several editions. His mission in life was to apply the methods of science to the study of clinical problems, a discipline he called "clinical science." Lewis was dedicated to promoting full time clinical research and, to this end, in 1930 he founded the Medical Research Society in the UK, which now has more than 1000 members. He also founded the journal *Clinical Science*.

Lillehei, C. Walton. A pioneering open-heart surgeon at the University of Minnesota in the mid 1950s, Lillehei introduced the practice of temporary cardiac pacing for patients who developed heart block as a result of surgical repair of congenital heart defects. At the end of the surgical procedure, Lillehei would insert the uninsulated tip of a wire lead into the myocardium and attach the other end to an external pulse generator.

Linde, Cecilia. Dr. Linde was born on August 8, 1950. She received her MD diploma in 1977 from the Karolinska Institute, Stockholm, Sweden. Her 1992 PhD thesis was entitled "Acute and long-term effects of atrioventricular synchronization in cardiac pacing." Presently, Linde holds the position of Associate Professor and Codirector of the Arrhythmia Unit in the Department of Cardiology, Thoracic Clinics at Karolinska Hospital in Stockholm. She became Director of the National Swedish Pacemaker Register in January 1996. Linde is a member of the Nucleus of the European Working Group on Cardiac Pacing (1998) and has been its Secretary since 2000. She has had responsibilities as a member of the editorial board of *Pacing and Clinical Electrophysiology: PACE* (1996), *Europace* (1998), and the *Mediterranean Journal of Pacing and Electrophysiology* (since 1998). She has chaired and co-chaired several international multicenter studies. Her research experience includes hemodynamics in cardiac pacing, women and arrhythmias, pacing in heart failure, and quality of life in pacemaker and cardiomyopathy patients.

Lindsay, Bruce D. Dr. Lindsay was born on September 16, 1951 in Kansas City, Missouri. His current professional address is Cardiovascular Division, Washington University School of Medicine, 660 South Euclid Avenue, Box 8086, St. Louis, MO 63110. Lindsay's present position is Associate Professor of Medicine and Director, Clinical Electrophysiology Laboratory, Washington University School of Medicine, St. Louis, MO. During his postgraduate training he was Intern (1977–1978) and Resident in Medicine (1978–1980) at the University of Michigan, Ann Arbor, Michigan. From 1980 to 1983 he was Medical Director, National Health Service Corps, East Jordan, Michigan; and, later on, a Fellow in Clinical Cardiology (1983–1985), Cardiology Division, Washington University School of Medicine, St. Louis, Missouri; and Visiting Pacemaker Fellow (6 months, 1986) at Newark Beth Israel Hospital, Newark, New Jersey. Lindsay was an Instructor in Medicine, Cardiology Division, Washington University School of Medicine, St. Louis, Missouri from 1985 to 1987. Since 1992 he has been an Associate Professor of Medicine, Cardiology Division, Washington University School of Medicine, St. Louis, Missouri. His honors and awards include Board of Trustees, American College of Cardiology (2003–2008), and Board of Trustees, North American Society of Pacing and Electrophysiology (NASPE) (2001–2006). Since 2003 Lindsay has been a member of the Executive Committee of the NASPE–Heart Rhythm Society. He is principle investigator of a study on electrophysiologic and physical determinants for successful abolition of ventricular tachycardia by endocardial catheter ablation. He has published numerous peer-reviewed papers in leading journals of cardiology, particularly on electrophysiology and pacing.

Lombardi, Federico. Dr. Lombardi was born in Brescia, Italy, on January 28, 1949. His post-MD training and specialization as a cardiologist were performed at the University of Milan, Italy. His medical training included a research fellowship in cardiology in the Lown Cardiovascular Group, Harvard Medical School and Harvard School of Public Health in 1980–1982 and a research fellowship in medicine at Brigham and Women's Hospital, Harvard Medical School. Since 2000 he has been an Associate Professor of Cardiology. Lombardi's main fields of interest include sudden cardiac death, ventricular arrhythmias, atrial fibrillation, acute myocardial infarction, autonomic control, heart rate variability, ventricular repolarization, antiarrhythmic drugs, and clinical electrophysiology. He is a fellow of the European Society of Cardiology (1994) and he received the H. C. Burger Award of the European Society of Noninvasive Cardiovascular Dynamics in 2001. Lombardi was a member of the Nucleus of the Working Group on Arrhythmias of the European Society of Cardiology. He is Italian Governor of the Board of International Society of Holter and Electrocardiology. He serves on several editorial boards, including the *Annals of Non-Invasive Electrocardiology, Cardiac Electrophysiology Review, European Heart Journal*, and *Italian Heart Journal*. Lombardi has published more than 500 scientific articles comprising 125 publications in major international cardiovascular journals, 98 chapters in books, and more than 290 abstracts.

Lown, Bernard. Dr. Lown was born on June 7, 1921. He emigrated from Lithuania to the USA. He is now an Emeritus Professor at Harvard School of Public Health and Senior Physician at the Brigham and Women's Hospital in Boston. He developed the DC defibrillator and the cardioverter, and introduced lidocaine as an antiarrhythmic drug. Lown and the electrical engineer Dr. Bernard Berkowitz studied the efficacy and safety of several DC waveforms in animals, showing that one was consistently effective in reversing the most intractable episodes of ventricular fibrillation that did not respond to alternating current. They learned that ventricular fibrillation could be prevented by synchronizing the shock to avoid the vulnerable period of the cardiac cycle, thereby providing a safe method for reverting tachycardias, a method that Lown called "cardioversion." His recent work demonstrates the role of psychological and behavioral factors in regulating the heart (see historical page 205).

Luceri, Richard M. Dr. Luceri was born in New York City. He completed his undergraduate education at the City University of New York. He then graduated from the University of Nancy, Faculté de Medecine, in Nancy, France. His postdoctoral training was received at Montefiore Medical Centre, Bronx, N.Y., where he had the privilege to be mentored in the arrhythmia device by Dr. Seymour Furman. This culminated in

Dr. Luceri's first of many published articles in the field. After completing his residency training in New York, his cardiology training was obtained at the University of Miami, where he was nominated to the Faculty as Assistant Professor of Medicine. In 1983, in collaboration with Dr. Michel Mirowski, he was instrumental in accomplishing the first implantation of an implantable cardioverter-defibrillator (ICD) in the southern USA. Subsequently, he continued this research after his move to Holy Cross Hospital in Fort Lauderdale in 1987, where he founded Florida Arrhythmia Consultants. He has been involved in the development and testing of many thousands of implantable arrhythmia devices of all types, and maintains one of the largest private-practice databases in the USA. He founded the *Journal of Electrophysiology* and was its Editor-in-chief from 1987 to 1990.

Lüderitz, Berndt. Dr. Lüderitz is Chairman of the Department of Medicine and Cardiology at the University of Bonn, Germany. His major interest is the management of tachyarrhythmias by pharmacologic and device therapy. During the era of shock-only implantable defibrillators, he associated a separate antitachycardia burst pacer with a defibrillator, creating a modern antitachycardia–implantable cardioverter-defibrillator (ICD) combination that led to the present combination of the two technologies. He has a special interest in the history of cardiology and the study of cardiac arrhythmias, about which he has written extensively. Lüderitz has received numerous awards and honors, including the Medal of Merit of the Polish Society of Cardiology (1997) and the Medal of Honor by the Council of the Medical Association of Cos—Island of Hippocrates (2000). He became a trustee of the North American Society of Pacing and Electrophysiology and an honorary member of the International Cardiac Pacing and Electrophysiology Society. In 2001, he was elected as an honorary member of the Slovenian Society of Cardiology. In November 2001, he obtained the Honorary Doctor of Medicine degree of the University of Athens, Greece.

Mahaim, Ivan. (*1897; †1965). Born on June 25, 1897 in Liège, Belgium, Mahaim was educated in Lausanne at the College Classique and started studying medicine in Lausanne in 1918. He was promoted to Doctor of Medicine in 1925. His medical career started at the Cantonal Hospital in Lausanne and continued at the Institute of Pathology. He was a fellow of Prof. Wenckebach in Vienna (1926) and of Prof. Clerc in Paris (1927). In 1959, he became an Associate Professor at the University of Lausanne. Mahaim wrote 100 papers that were published in all the important journals of his time. His most influential works were his books on histologic research concerning the connections of the bundle of His, which

Ivan Mahaim

was a resounding success in Europe in 1937 as it provided the basis for later electrophysiologic discoveries. Another of his publications, which is used as a reference today, is his book *Les Maladies Organiques du Faisceau de His—Tawara*, which is devoted to the tumors of the heart, providing a detailed analysis of more than 400 cases. This book anticipates the possibility of cardiac surgery, which at that time was considered science fiction. His last big *oeuvre* was devoted to Beethoven and published in 1964. It is still a reference for musicians, as it is the most detailed analysis and description of the background on Beethoven's last quatuors. With these fine works, Mahaim proved his outstanding personality as a detailed analyst, a great artist, and a devoted writer. In-depth study of the work of Mahaim will certainly lead one to an outstanding human being, to a pioneering time of medicine and cardiology, to the harmony between the first concepts of medicine and cardiology, and to the harmony between the first concepts of electrophysiology and the quatuors of Beethoven (see historical page 168).

Malik, Marek. Dr. Malik was born in Prague, in Czechoslovakia (now known as the Czech Republic), on April 23, 1951. His present position is Professor of Cardiac Electrophysiology at the Department of Cardiological Science at St. George's Hospital Medical School in London, UK. He has worked at Charles University, Faculty of Mathematics and Physics in Prague, most recently as Professor of Computer Science (1974–1987).

From 1978 to 1987 he was a consultant in the Department of Internal Medicine of the Charles University, University Hospital. Malik held many offices in scientific societies dealing with computers in cardiology, Holter and non-invasive electrocardiology, and databases and telecommunications. He has served on the editorial boards of several major journals of cardiology, including *Pacing and Clinical Electrophysiology: PACE, Clinical Cardiology, Annals of Non-invasive Electrocardiology, Cardiac Electrophysiology Reviews, Journal of the American College of Cardiology*, and, since 2001, *The Journal of Cardiovascular Electrophysiology*. He has published more than 260 scientific articles in peer-reviewed journals and is the author of nine books and textbooks. He is also a fellow of the European Society of Cardiology and a fellow of the American College of Cardiology.

Manolis, Antonis S. Dr. Manolis was born on July 12, 1954 in Kastania-Nafpaktias, Greece. His last position was Professor and Director of Cardiology at Patras University Medical School, Rio, Patras, Greece. He was educated at the Athens University Medical School, Athens, Greece, the Cabrini Medical Center, the New York Medical College, and St. Vincent's Medical Center, all in New York. Manolis completed an advanced fellowship in cardiac electrophysiology and pacing at Tufts University/New England Medical Center in Boston. He became a Professor and Director of Cardiology at Patras University Medical School, Greece in 1995. He is President of the Working Group of Cardiac Pacing and Electrophysiology (Hellenic Society of Cardiology). As a scientist, Manolis holds many important offices and memberships on the editorial boards of leading publications. He has to his credit many scientific publications in international peer-reviewed medical journals and in books.

Marchlinski, Francis E. Dr. Marchlinski was born in Nanticoke, Pennsylvania, on June 23, 1951. He received his MD degree in 1976 from the University of Pennsylvania Medical School in Philadelphia. His present position is Professor of Medicine at the University of Pennsylvania School of Medicine and Director of Cardiac Electrophysiology in the University of Pennsylvania Health System. Marchlinski has received numerous honors and awards, including the Dr. Morris Ginzburg Award in 1976, the Osler Award from the Master University of Miami in 1990, and the Presbyterian Medical Center House Staff Teaching Award in 1993. He has worldwide editorial responsibilities and has served, and still serves, on many editorial boards, including those of *The Journal of Cardiovascular Electrophysiology, The American Journal of Cardiology, Journal of the American College of Cardiology*, and *Pacing and Clinical Electrophysiology: PACE*. Marchlinski has been course director of several conferences on electrophysiology and pacing. He has published significant articles in the major English-language journals of cardiology.

Marcus, Frank I. Dr. Marcus was born on March 23, 1928, Haverstraw, New York. He is now an Emeritus Professor of Medicine in the Section of Cardiology of the Department of Medicine at the University of Arizona Health Sciences Center in Tucson, Arizona. He received his education at Columbia University in New York City, and at Tufts University and Boston University in Boston. His postgraduate education took place at the Peter Bent Brigham Hospital in Boston and the Georgetown University Hospital in Washington, D.C. From mid 1982 to 1999 he was a Distinguished Professor of Medicine, Director of Arrhythmia Services, and Director of the Pacemaker Clinic at the University of Arizona. Marcus has received numerous honors and awards, including the Award in Excellence, Cardiology Section, University of Arizona College of Medicine in 1981 and the Outstanding Reviewer Award, *The American Journal of Cardiology*, in 1995. He has served on the editorial and scientific boards of major English-language journals, including *Circulation* and *Journal of the American College of Cardiology*, and as Associate Editor of *The American Journal of Cardiology*. His scientific interests include clinical electrophysiology, particularly ablation of supraventricular arrhythmias, arrhythmogenic right ventricular dysplasia, and the molecular and genetic basis of arrhythmias.

Marey, Etienne Jules. (*1830; †1904). E. J. Marey was born March 5, 1830 in Beaune, Burgundy (Côte d'Or), France. His father was an assistant to a Burgundy wine merchant. Marey's thesis, completed at the age of 29 years, was dealing with the circulation of the blood under normal and pathological conditions. This work marks the beginning of his studies on cardiovascular hemodynamics. His main interest, however, was engineering, particularly focused on graphic registration of changes in time and place; especially concerning movement of the heart and pulsation of the arteries or locomotion of animals—in the air, in the water, and on the ground—and humans. Marey's interest centered around movement in all of its forms: cardiovascular hemodynamics, respiration, muscular contraction, and complex movements. For this he invented many scientific instruments himself. The graphic registration was replaced after 1881 by photographic records. Marey contributed considerably to the development of this new tool of registration and evaluation of movement ("chronophotography"), which finally resulted in the invention of the cinematograph. Furthermore, Marey's studies on the flight of birds lead him to research on gliding and aviation. Marey, who was a professor at the Collège de France and a member of the Academy of Science, produced numerous scientific papers, drawings, photographs and films (the first in the history of cinema!). Marey, a technical genius, the "engineer of life" and inventor of cinematography died on May 15, 1904 at his Paris home (see historical page 144).

Mason, Jay W. Mason was born on July 22, 1946 in St. Louis, Missouri, USA. His credentials are: Princeton University, BA; University of Pennsylvania School of Medicine, MD. His current position is Medical Director, Covance, Central Diagnostics, and Chair, Cardiovascular Sciences Group. Mason's past positions were Jack Gill Professor and Chair, Department of Internal Medicine, University of Kentucky, College of Medicine; Chief, Division of Cardiology, University of Utah, School of Medicine; and Director, Cardiac Arrhythmia Study Unit, Stanford University School of Medicine. He has authored pivotal papers on cardiac arrhythmias in the leading journals of the fields. He has served as invited speaker at national and international scientific conferences on cardiac arrhythmias. Mason's stress busters are wine collecting and fly-fishing. He is married to Molly and has three children, one son and two daughters.

Meinertz, Thomas. Dr. Meinertz, born on August 14, 1944 in Warendorf, Germany began his medical education at the University of Mainz in Germany. After receiving his MD degree, Meinertz continued at the University of Mainz until he achieved the position of Associate Professor at the University of Freiburg. He was appointed Professor of Cardiology and Medicine at the University of Hamburg in 1994, and currently holds the position of Chief of the Department of Cardiology at the University Hospital, Eppendorf, Hamburg, Germany. Meinertz has authored numerous clinical papers and abstracts in the field of cardiology and is currently President of the German Cardiac Society. He has been a member of the editorial board of *Circulation* since 1999.

Mirowski, Mieczyslaw (Michel). (*October 14, 1924, Warsaw, Poland; †March 26, 1990). Between 1939 and 1945, Dr. Mirowski, born Mordechai Friedman, survived Nazi-controlled Poland, and finally escaped. He went to France, where he studied medicine and graduated in 1953. Mirowski's postgraduate training took place in Israel, Mexico, and the USA (Baltimore, Maryland). After 5 more years in Israel as a cardiologist, he moved back to Baltimore. In 1969, together with Dr. Morton M. Mower, he began work on designing the defibrillator and managing associated financial and technical problems. In 1975, after extensive animal studies, a prototype was fully implanted in a dog. The device was soon ready for use in humans, and on February 4, 1980, Mirowski

'Michel' Mirowski

successfully implanted the first automatic defibrillator in a human at the Johns Hopkins Hospital in Baltimore. Thereafter, the automatic implantable defibrillator and, later on, the automatic implantable cardioverter-defibrillator (ICD), became one of the most successful therapeutic tools in cardiology.

Mitsui, Toshio. Dr. Mitsui, a cardiac surgeon, was early on involved with the management of cardiac arrhythmias by external DC counter-shock and cardiac pacing. In 1968, he delivered an important presentation called "Optimal heart rate in cardiac pacing in coronary sclerosis and non-sclerosis" at the Second World Symposium on Cardiac Pacing. In this presentation he coined the term "pacemaker syndrome." Mitsui has been President of the Japanese Society of Artificial Organs and is Provost of the University of Tsukuba School of Medicine.

Mobitz, Woldemar. (*May 31, 1889, St. Petersburg, Russia; †April 11, 1951, Freiburg, Germany). Dr. Mobitz was a Professor of Internal Medicine in Freiburg. In 1924, he became an Associate Professor in Munich. Later he worked in Freiburg as a Senior Consultant at the Medical University Clinic, and was then Head of the Medical Clinic of the Municipal Hospital in Magdeburg until the occupation by the Soviet army in 1945. Mobitz' name became known through his fundamental work on atrioventricular (AV) conduction blocks and their classification

(1924), as well as the future designation "AV-block, II Grade, Type Mobitz." (See historical page 155.)

(Z. ges exp Med 41 (1924) 180–237)

Über die unvollständige Störung der Erregungs-überleitung zwischen Vorhof und Kammer des menschlichen Herzens.

Von

W. Mobitz.

(Aus der I. medizinischen Klinik zu München [Direktor: Prof. *von* Romberg].)
Mit 26 Textabbildungen.
(Eingegangen am 29. Dezember 1923.)

Seit den Untersuchungen von *His jun.* (1893) wissen wir, daß ein histologisch den übrigen Muskelfasern des Herzens sehr nahestehendes Faserbündel die bindegwebige Grenze zwischen Vorhöfen und Kammern des Warmblüterherzens an einer bestimmten Stelle durchbricht und die Muskulatur beider Herzabteilungen verbindet. *Aschoff* und sein Schüler *Tawara* fanden in dem den Vorhöfen

Moe, Gordon K. (*May 30, 1915, Fairchild, Wisconsin; †October 24, 1989). Dr. Moe received his PhD degree in 1940 from the University of Minnesota and his MD degree 1943 from Harvard Medical School. In 1946, he became an Associate Professor of Pharmacology at the University of Michigan, and from 1950 to 1960 he was Professor of Physiology and Chairman of the same department at the State University of

Gordon K. Moe

New York College of Medicine, Syracuse. From 1960 until his retirement in 1984, Moe was the Director of Research at the Masonic Medical Research Laboratory in Utica, New York. He was an honorary fellow of the American College of Cardiology and has received numerous awards and honors. His many important publications are devoted to cardiac arrhythmias, particularly the dual atrioventricular (AV) transmission system, atrial flutter and fibrillation, and the multiple wavelet hypothesis. He was interested in computer and mathematical models of cardiac arrhythmias, concealed conduction, and many other pathophysiological and pharmacological aspects of cardiac arrhythmias. He was one of the true great individuals in electrophysiology and arrhythmology.

Mond, Harry George. Dr. Mond was born in Melbourne, Australia, on December 20, 1943. He received his MD degree in 1973. His doctoral thesis at the University of Melbourne was entitled "Assessment of myocardial function in acute myocardial infarction." Mond became a Fellow in Cardiology in 1974 at Emory University in Atlanta, Georgia. His present position is Senior Medical Staff and Physician to Pacemaker Clinic at the Royal Melbourne Hospital, Associate Professor in the Department of Medicine at the University of Melbourne, and Honorary Pacemaker Specialist at Royal Children's Hospital in Victoria. He is a founding member of NASPExAM (North American Society of Pacing and ElectrophysiologyxAM) in 1985. Mond is a board member of the International Pacing and Electrophysiology Society and the Asian–Pacific Society of Pacing and Electrophysiology. He is coordinator of the World Survey of Cardiac Pacing and ICD and serves as Editor of *Asia–PACE News*. He is also a fellow of the Royal Australian College of Physicians and a fellow of the American College of Cardiology.

Morady, Fred. Dr. Morady was born in Paris, France; he immigrated to the USA with his parents when he was 4 years old. Morady attended Medical School at the University of California, San Francisco, and stayed there for his internship, residency, and cardiology fellowship, benefiting from superb teachers such as William Parmley, Kanu Chatterjee, and Melvin Scheinman. In 1984, Morady moved to Ann Arbor to become Director of the Clinical Electrophysiological Laboratory at the University of Michigan. He has served as the "Arrhythmia of the month" Section Editor for the *Journal of Cardiovascular Electrophysiology* over the past several years. His scientific creativity has resulted in over 400 peer-reviewed original contributions. Morady has been married to Paulette Metoyer, MD, for 24 years, and has two daughters. Morady received at Heart Rhythm 2004 the North American Society of Pacing and Electrophysiology's Distinguished Teacher Award.

Morgagni, Giovanni Battista. (*February 25, 1682, Forli, Italy; †December 5, 1771). Founder of pathological anatomy. Morgagni began his study of medicine and philosophy in 1698 at the University of Bologna, Italy. He earned his doctorate in medicine and philosophy in 1701 and spent a few years working for Valsalva, his former teacher (1666–1723), whom he succeeded in 1706 in the field of anatomy and clinical medicine. In 1711, after 4 years of practice, Morgagni became a Professor of Theoretical Medicine at the University of Padua, Italy, and also in 1711 a Professor of Anatomy, a position he held until his death. In 1761 Morgagni published his principal work *De Sedibus et Causis Morborum per Anatomen Indagatis*, in which he compiled several hundred case studies in the form of letters. In the 64th letter of this work, *Ad Thoracis Morbos Pertinet*, Morgagni presents a clinical description of circulatory arrest in a patient, in the course of which he implies a relationship between a slow pulse and a syncopal attack. Morgagni describes bradycardia, convulsive reactions of the vasomotor reactions, and the phenomenon of facial coloration following the attack. In this context, he understood the connection between circulation and bradycardia or extreme tachycardia.

G.B. Morgagni

Arthur Moss

Moss, Arthur. Dr. Moss was born in 1931 in White Plains, New York, and was educated at Yale University and Harvard Medical School. He has been a Professor of Medicine at the University of Rochester Medical Center since 1991. He was been involved in the introduction of transvenous left atrial pacing for the control of recurrent ventricular fibrillation, the identification of pacemaker-induced extracardiac sounds, and permanent pervenous atrial synchronized ventricular pacing. He was also the first to utilize implanted pervenous pacemakers to inhibit and terminate recurrent ventricular tachycardia. His work on the congenital long QT syndrome (LQTS) spans three decades. He successfully performed the first left cardiac sympathetic denervation in 1969. He played a fundamental role in his collaboration with the molecular biologists that successfully identified the first LQTS genes. More recently he has evaluated the implantable cardioverter-defibrillator in MADIT I, II and in subsequent studies that have demonstrated its benefit toward survival.

Mower, Morton. Dr. Mower was born in Baltimore, Maryland, in 1933. He attended Johns Hopkins University and obtained his MD degree at the University of Maryland School of Medicine. He is the coinventor of the automatic implantable cardioverter-defibrillator. He holds 21 patents as a coworker of Dr. Mirowski. The concept of an implantable device that would at once recognize and defibrillate ventricular fibrillation has revolutionized the care of cardiac arrest patients, and has saved thousands of lives over the past 20 years. Its impact continues to grow as we recognize new groups of high-risk patients that need prophylactic defibrillators. Mower has recently reinvented his visionary career with innovative investigations into the role of biventricular pacing in congestive heart failure.

Mugica, Jacques Edmond. (1933–2002). Dr. Mugica, a surgeon, founded and headed a surgical and then cardiologic pacemaker–implantable cardioverter-defibrillator (ICD) clinic in Paris that has seen perhaps the largest group of pacemaker implants in the world. In addition, he began Cardiostim, a series of highly influential and scientifically advanced symposia conferences that are held every 2 years in Nice, France. Mugica was the recipient of the 1995 Distinguished Service Award of the North American Society of Pacing and Electrophysiology (see historical page 180).

Murgatroyd, Francis David. Dr. Murgatroyd was born on March 11, 1961 in London, UK. His current position is Consultant Cardiologist, and his professional address is Cardiac Electrophysiology Unit, Papworth Hospital, Cambridge CB3 8RE, UK. His postgraduate training was completed at Guy's Hospital, London (1986–1987) and St. George's Hospital, London (1987–1989). His specialist (cardiology) training was performed at St. George's Hospital, London (1989–1995) and Glenfield Hospital, Leicester (1996–2000). His subspecialist (electrophysiology) training took place at St. George's Hospital, Medical School, London (1990–1995), the University of Western Ontario, Canada (1998–1999), and Glenfield Hospital, Leicester, UK (1999–2000). His publication activity includes 45 papers, three books, 19 book chapters, and more than 100 published abstracts. His principal interest is antiarrhythmic therapies for atrial fibrillation and flutter: conducted multicenter drug trials. He pioneered the clinical use of internal cardioversion device-based treatments and catheter ablation. Murgatroyd published *Atrial Fibrillation for the Clinician* (Futura, 1995), *Non-Pharmacological Management of Atrial Fibrillation* (Futura, 1997), and *Handbook of Cardiac Electrophysiology* (Remedica, 2002). His clinical interest concerns clinical electrophysiology and catheter ablation, defibrillators, and pacemakers.

Myerburg, Robert J. Dr. Myerburg received his undergraduate and medical training at Johns Hopkins University and the University of Maryland, respectively. In 1970, Myerburg was appointed Assistant Professor of Medicine at the University of Miami, followed a year later by the same appointment in the Department of Physiology. He currently serves as Director of Cardiology at the University of Miami, a position he has held since 1973. Myerburg has explored ion channels, cellular electrophysiology, organ electrophysiology, and human cardiac electrophysiology. He is also familiar with implanted pacemakers and defibrillators, as well as with public health issues. His contributions in the field of sudden cardiac death are outstanding, from his early explorations into the electrophysiology of the specialized conducting tissue that demonstrated the "gates of Myerburg" (lengthening of action potential duration and refractoriness along the bundle branches until just prior to insertion into ventricular muscle) to his longstanding ongoing interest in cardiac electrophysiology after healed myocardial infarction. Myerburg has published a rich database of electrophysiology. He was honored by the North American Society of Pacing and Electrophysiology with the 2000 Distinguished Scientist Award, and by the University of Miami with the 2001 Distinguished Faculty Scholar Award.

Naccarelli, Gerald V. Dr. Naccarelli is a Professor of Medicine, Chief of the Division of Cardiology, and Director of the Cardiovascular Center at the Pennsylvania State University College of Medicine/ Milton S. Hershey Medical Center. He received his MD degree from the Pennsylvania State University College of Medicine and continued

Gerald V. Naccarelli

his internal medicine training at the North Carolina Baptist Hospital/ Bowman Gray School of Medicine. Prior to his current position at Hershey, he was Director of Clinical Electrophysiology and Vice Chairman of Cardiology at the University of Texas Medical School at Houston. He has written two books and he serves on the editorial boards of *The American Journal of Cardiology*, *The Journal of Cardiac Electrophysiology*, *Clinical Cardiology*, *Journal of Interventional Electrophysiology*, *Journal of Cardiovascular Pharmacology and Therapeutics*, and *Pacing and Clinical Electrophysiology: PACE.*

Natale, Andrea. Dr. Natale was born in Sircusa, Italy. He graduated *summa cum laude* from the Medical School of the University of Firenze, Italy in 1985 and *summa cum laude* from the Catholic University School of Cardiology in Rome, Italy. Natale received his clinical training in cardiology at the Methodist Hospital, Baylor College in Houston, Texas, and at the University of Western Ontario, London, Ontario, Canada. Natale is a board certified staff cardiologist in the Cleveland Clinic Heart Center who specializes in the treatment of abnormal heart rhythms. He is the Cosection Head of Pacing and Electrophysiology and Director of the Electrophysiology Laboratories in the Department of Cardiovascular Medicine. Natale has a special interest in the treatment of atrial fibrillation. He has pioneered some of the present catheter-based cures for atrial fibrillation. He was also one of the first electrophysiologists to perform percutaneous epicardial radiofrequency ablation, which is a treatment for patients who fail conventional ablations. Natale has held several academic appointments including Head of the Cardiovascular Physiopathology Section, Italian Air Force, Aerospace Research Center, Department of Medicine, Rome Italy; Assistant Professor of Medicine, Director of Electrophysiology Laboratory, Duke University, Durham, North Carolina; and Associate Professor of Medicine, Director of Electrophysiology Program, University of Kentucky, Lexington. He currently serves as a Professor of Medicine, Ohio State University. Natale has been an invited lecturer at more than 50 symposiums and conferences worldwide. He is the author or coauthor of more than 200 published studies on pacing and electrophysiology. Natale serves as a reviewer on numerous editorial boards of the leading journals in the field. In his leisure time, he enjoys skiing and sailing.

Nattel, Stanley. Dr. Nattel was born in Haifa, Israel, on January 28, 1951. His present position is Director of the Research Center at the Montreal Heart Institute in Montreal, Quebec, Canada. He received his postgraduate training at the Royal Victoria Hospital, Montreal General Hospital, Indiana University, and the University of Pennsylvania. Among

Nattel's most important scientific contributions are the first demonstration of the Ca^{2+} channel blocking actions of amiodarone and its effect on the atrioventricular (AV) node, a detailed elucidation of ionic currents governing repolarization in the human atrium, an explanation of the ionic mechanism of atrial tachycardia remodeling and mechanisms of atrial fibrillation associated with congestive heart failure, and the development of novel approaches to inhibiting the development of atrial fibrillation substrate. Nattel has received numerous honors and awards. He has served on the editorial boards of many major English-language pharmacology and cardiology journals, including *Journal of Cardiovascular Pharmacology, Circulation Research,* and *Journal of the American College of Cardiology.*

Olsson, Bertil S. Dr. Olsson was born in Katrineholm, Sweden, on May 10, 1941. His present position is Professor, Department of Cardiology at the University Hospital in Lund, Sweden. He received his MD diploma from Gothenburg University in Sweden in 1965. His thesis was entitled "Monophasic action potentials of right heart suction electrode methods in clinical investigations." He became an Associate Professor of Medicine at Gothenburg University in 1972. Olsson has published 206 scientific articles in major peer-reviewed cardiology journals.

Oseroff, Oscar. Dr. Oseroff was born in Neuquén, Argentina, on February 1, 1948. He graduated with a degree in medicine (MD) in March 1972 from Cordoba University in Argentina. In 1977, after his cardiology residency (1973–1976) at the Castex Hospital in Buenos Aires, Argentina, he became Chief of the Electrophysiology and Pacing Department at Castex Hospital. Oseroff completed a research fellowship at the Montefiore Medical Center in the Bronx, New York, from 1986 to 1987. He was President of the Scientific Committee of the Tenth International Symposium on Cardiac Pacing and Electrophysiology in Buenos Aires in 1995. Oseroff has presented more than 140 papers at major national and international meetings on pacing and electrophysiology. He is a member of the Advisory Board of the North American Society of Pacing and Electrophysiology, a member of the Board of Directors of the International Society of Pacing and Electrophysiology (ICPES), Vice President of SOLAECE (Latin American Society of Pacing and Electrophysiology), and Secretary of the Cardiac Arrhythmia Board, Argentinean Society of Cardiology.

Oto, Ali. Dr. Oto was born on January 5, 1954 in Diyarbakir, Turkey. He graduated with honors in 1976, and became a Professor of Medicine and Cardiology in 1990. He served as Dean of Inönü University School of

Medicine, Turkey, and was Acting President of the University from 1990 to 1992. His present position is Professor of Medicine and Cardiology in the Department of Cardiology at Hacettepe University School of Medicine in Ankara, Turkey. Oto has published several scientific books on pacing and electrophysiology. He has organized very successful congresses in Turkey, including the Seventh European Symposium on Cardiac Pacing in Istanbul (1995) and the Congress of the International Society for Holter and Non-Invasive Electrophysiology and ICPEP in Istanbul (2000). Oto has several editorial responsibilities. His most recent book is called *Myocardial Repolarization: From Gene to Bedside*.

Ovsyshcher, I. Eli. Dr. Ovsyshcher was born in the former USSR on May 30, 1936. He received his MD degree in 1960 in St. Petersburg, Russia, and his PhD in 1971. He immigrated to Israel in 1973. He made use of a fellowship in invasive cardiology at Soroka University Medical Center in Beer-Sheva, Israel from 1973 to 1975, and a fellowship in electrophysiology at Stanford University Medical School in California in 1983. Ovsyshcher's present position is Professor of Medicine/Cardiology and Director of the Arrhythmia Services and Research Electrophysiology Laboratory, Cardiology Department, Soroka University Medical Center and Faculty of Health Sciences, Ben Gurion University of the Negev in Beer-Sheva, Israel. Ovsyshcher is Chairman of the Israel Working Group on Cardiac Pacing and Electrophysiology. He is well known as founder and President of the International Dead Sea Symposia on Cardiac Arrhythmias and Device Therapy. In 1985, Ovsyshcher performed the first atrioventricular (AV) node ablation in Israel. He serves on numerous editorial boards of highly renowned medical journals. He is principle investigator of more than 20 research projects, and has published more than 270 scientific papers. He is also a fellow of the European Society of Cardiology and a fellow of the American College of Cardiology.

Pacifico, Antonio. Dr. Pacifico was born on August 16, 1950 in Italy. He is Clinical Associate Professor of Medicine, Baylor College of Medicine and Chairman, Texas Arrhythmia Institute. He was President of the Houston Cardiology Society (1998). In 1989, he became President of the Houston Electrophysiology Society. In 2002, he was appointed as Editor-in-chief of *Implantable Defibrillator Therapy: A Clinical Guide*. His main interest concerns are: "Design and implementation of a new computerized system for intraoperative cardiac mapping: experience during surgery" (*Journal of Applied Physiology*, 1991); "Long-term follow-up of cardioverter-defibrillator implanted under conscious sedation in prepectoral subfascial position" (*Circulation*, 1997); and "Prevention of implantable defibrillator shocks by treatment with sotalol" (*New England Journal of Medicine*, 1999).

Packer, Douglas L. Dr. Packer was born on October 12, 1953 in Utah. He completed his education in medicine at the University of Utah College of Medicine in 1980. His present position is Professor in Medicine at the Mayo School of Medicine in Rochester, Minnesota. Packer has received numerous honors, including the 1983 Haskel Schiff Award in Internal Medicine. He has held many important national and local appointments and has received numerous grants and support for innovative investigations, recently for pulmonary vein ablation studies. Packer has been a reviewer for major journals including *Circulation, Circulation Research, The American Journal of Cardiology,* and *Pacing and Clinical Electrophysiology: PACE.* He serves on the editorial boards of *Journal of the American College of Cardiology* and *American Heart Journal,* and he is Associate Editor of *The Journal of Cardiovascular Electrophysiology.* He regularly lectures internationally on a variety of topics having to do with cardiac arrhythmias and electrophysiology.

Pappone, Carlo. Dr. Pappone was born on December 5, 1961 in Benevento, Italy. His current position is Director, Division of Arrhythmology, Department of Cardiology, Hospital San Raffaele, Milan, Italy. His fields of expertise cover electrophysiology, catheter ablation of cardiac arrhythmias, ablation of atrial fibrillation, non-fluoroscopic mapping, cardiac pacing, biventricular pacing, and non-excitatory cardiac contractility modulation. Pappone received his MD degree on October 29, 1986 from the School of Medicine, University of Naples Federico II., in Italy, with honors. In the same year he received his medical licensure, with honors, and in October 1995 received his PhD degree, with honors. His thesis was entitled "Transcatheter radiofrequency ablation of atrial tachycardia: mapping techniques." His academic and research appointments include the position of Associate Professor, Interventional Electrophysiology, Cardiology Fellowship Program, University of Naples Federico II., and Director of the Course on Cardiovascular Diseases, S. Raffaele University Medical School, Milan (1999–2000). In 2002, he was Invited Visiting Professor at the University of Michigan, Division of Cardiology, Ann Arbor, Michigan. Pappone's international scientific reputation is based on landmark studies on radiofrequency transcatheter ablation for atrial fibrillation published for 3 consecutive years in *Circulation.* He authored a leading study entitled "Catheter ablation of paroxysmal atrial fibrillation using a 3D mapping system (Circulation, 1999)." Pappone conducted seminal research on the electrical treatment of patients with heart failure, including cardiac resynchronization therapy using multisite pacing and the novel non-excitatory electrical cardiac contractility modulation signal. Pappone has received several honors and awards, including the

Honorary Membership of the Russian Society for Cardiac Pacing and Electrophysiology.

Parsonnet, Victor. Surgeon Dr. Parsonnet of Newark Beth Israel Medical Center began his contributions with hospital-based intraoperative cardiac monitoring in 1952. In 1961, he introduced a temporary transvenous bipolar pacing lead that has been in use for emergency and temporary pacing ever since. He later devised the subcostal incision for pacemaker and implantable cardioverter-defibrillator (ICD) implantation, and implanted the first DDD pacemaker. He was codeveloper of the introducer technique for pacemaker implantation, and, in 1988, with his associates, was the first to implant a transvenous ICD. Parsonnet is a founding member of the International Cardiac Pacing and Electrophysiology Society, which has conducted quadrennial world symposia since 1963. Parsonnet was a cofounder of the North American Society of Pacing and Electrophysiology and served as its President 1990–1991.

Pasteur, Louis. (*December 27, 1822, Dole, Jura/France; †September 28, 1895, Villeneuve-l'Étang, near Paris, France). Chemist and founder of

microbiology. With his detection of microbes, his method of killing them (pasteurization), and the development of protective immunizations, Pasteur led medicine into the age of bacteriology: He is famously known for saying "In the area of observation, chance only benefits the mind that is prepared!" (Pasteur). Pasteur was a former Honorary Doctor of the University of Bonn, Germany; however, as a Frenchman, Pasteur returned his honorary doctorate after the bombardment of Paris by German troops in 1871. Nevertheless, he once wrote that science knows no single native country—"*la science n'a pas de patrie.*"

Petrac, Dubravko. Dr. Petrac was born on March 14, 1944 in Zagreb, Croatia. He received his MD at the University of Zagreb Medical School in 1971. He became a specialist in internal medicine in 1978, and finished his postgraduate study of cardiology at the University of Zagreb in 1983. He improved his skills in electrophysiology with Prof. P. Touboul in Lyon, France, in 1983–1984. In 1993, he became a Doctor of Medical Science, and in 1998 he became a Fellow of the European Society of Cardiology. In 2002, he became Professor of Internal Medicine at the University of Zagreb. He is currently Chief of the Department of Cardiology, Center of Cardiac Pacing and Laboratory of Cardiac Electrophysiology at the Sestre Milosrdnice University Hospital in Zagreb. Petrac is one of the founders of the Alpe Adria Association of Cardiology and the Mediterranean Society for Cardiac Pacing and Electrophysiology. He has published over 150 scientific and professional papers and four books on cardiac arrhythmias.

Pick, Alfred. (*1907, Prague, in what is now known as the Czech Republic; †January 8, 1982, Chicago, Illinois). Dr. Pick was a distinguished electrocardiographer. He received his MD degree in Prague in 1932. In 1949, he moved to the USA and began an association with the Michael Reese Hospital and Medical Center in Chicago, where he worked for the rest of his career. Pick wrote definitive papers on aberrant conduction, reciprocal beating, parasystole, preexcitation, digitalis intoxication, the supernormal phase of atrioventricular conduction, and the diagnosis of supraventricular versus ventricular tachycardias. With his long time collaborator Dr. Richard Langendorf, he published the text *Interpretation of Complex Arrhythmias*. Pick served on the editorial boards of the *American Heart Journal* and *Journal of Electrocardiology*, and was a member of the International Scientific Board of Coeur et Medicine Interne (Paris, France). He was a Professor of Medicine at the Pritzker School of Medicine of the University of Chicago and Senior Consultant to the Michael Reese Hospital and Medical Center when he passed away.

Porstmann, W. (*1921; †1982). Dr. Porstmann developed the first technique for the percutaneous occlusion of the duct of Botallo that he introduced in 1966 and successfully applied in 139 patients until 1978. As a result of his close personal friendship with Dotter, Porstmann performed the first peripheral angioplasties in Europe at the Charité Medical School in the former German Democratic Republic ("East Germany").

Pratt, Craig M. Dr. Pratt is a Professor of Medicine and Director of Clinical Cardiology Research at Baylor College of Medicine in Houston, Texas. He is also Director of the Coronary Intensive Care Unit at the Methodist Hospital and Director of Outpatient Cardiovascular Services for the MacGregor Medical Association, Houston. Pratt is a past Chairman of the Cardiovascular and Renal Drugs Advisory Board of the Food and Drug Administration and continues to serve as a consultant to the Center for Drug Evaluation and Research. His primary area of research is the study and development of new therapies for the treatment of arrhythmias, myocardial infarction, ischemia, and heart failure. He has been a principal investigator for a number of pivotal studies, including several National Heart, Lung and Blood Institute (NHLBI) supported projects.

Priori, Silvia G. Dr. Priori received her education at Liceo Scientifico, Vigevano and the University of Milan in Italy. Her professional experience includes relevant positions at the Universities of Pavia and Milan, and also a research fellowship at Washington University in St. Louis, Missouri. Priori has twice received the Young Investigator Award from the Italian Society of Cardiology (1990, 1995). Her bibliography of 320 publications is devoted to molecular and genetic electrophysiology and the genetic and molecular basis of cardiac arrhythmias and sudden cardiac death. Priori is Chairman of the Task Force on Sudden Cardiac Death of the European Society of Cardiology.

Pritchett, Edward L.C. Dr. Pritchett is Professor of Medicine in the Divisions of Cardiology and Clinical Pharmacology at Duke University Medical Center in Durham, North Carolina. He has published extensively on supraventricular arrhythmias and preexcitation syndromes and on the clinical pharmacology of drugs used to treat these disorders. He has served as Chief of the Division of Clinical Pharmacology and as Director of the General Clinical Research Center at the Duke University Medical Center. Pritchett has been a member of the Cardiovascular and Renal Drugs Advisory Committee for the US Food and Drug Administration, the General Research Center's Advisory Committee for the National Institutes of Health, and the Cardiovascular Physiology and Pathophysiology Research Study Committee for the American Heart Association.

Eric N. Prystowsky

Prystowsky, Eric N. In 1973 Dr. Prystowsky graduated from the Mount Sinai School of Medicine in New York, where he interned in medicine from 1973 to 1974, and subsequently served his residency in medicine from 1974 to 1976. From 1976 to 1979 he was a cardiology fellow at Duke University Medical Center in Durham, North Carolina. From 1979 to 1988 he was a member of the teaching faculty at Indiana University School of Medicine, where he was an Associate Professor of Medicine, and at Duke University Medical Center, where he was a Professor of Medicine. Prystowsky is currently Director of the Clinical Electrophysiology Laboratory at St. Vincent Hospital and a Consulting Professor of Medicine at Duke University Medical Center. He was President of the North American Society of Pacing and Electrophysiology in 2002.

Puech, Paul. Dr. Puech was Professor and Head of the Department of Cardiology at the University of Montpellier in France. He was a specialist of arrhythmias and published many papers on the diagnosis and management of different arrhythmias. He was the first to directly record the electrical activity of the His bundle. Following this initial recording in 1960, he pursued, with his collaborator Dr. Robert Grolleau, a series of investigations concerning His bundle activity in different cardiac diseases. He is considered one of the pioneers of modern invasive electrophysiology. While he was President of the French Society of Cardiology 1979–1980, Puech was also President of the Congress of the European Society of Cardiology held in Paris, June 22–26, 1980. He was also

President of the International Society and Federation of Cardiology from 1985 to 1986.

Raviele, Antonio. Dr. Raviele, born August 26, 1946, graduated with honors in medicine and surgery from Naples University, Italy, in 1971, and then became a Specialist in Cardiovascular and Rheumatic Diseases at the University of Florence, Italy. Since 1971 he has been employed as a hospital physician in cardiology departments. Since 1995 he has also been the Director of the Cardiology Operative Unit of Umberto I., Hospital of Mestre/Venice. He is currently Chief of the Cardiovascular Department at the same hospital. Raviele is the author of more than 600 scientific publications in international peer-reviewed journals. He has edited and supervised production of several books and has been the promoter, organizer, and coordinator of numerous scientific meetings in the field of arrhythmology. He is President and Scientific Secretary of the "International Workshop on Cardiac Arrhythmias," which has been held every 2 years since 1989 in Venice and is sponsored by the European Society of Cardiology, the North American Society of Pacing and Electrophysiology, the International Society for Holter and Noninvasive Electrocardiology, and the World Heart Federation. Raviele is a founding member and coordinator of the Italian Arrhythmology Group (GIA), and has been President of the Italian Association of Arrhythmology and Cardiac Pacing (2000).

Reiffel, James A. Dr. Reiffel received his undergraduate training at Duke University and his medical degree from Columbia University College of Physicians and Surgeons. He is a faculty member of the Clinical Electrophysiology and Pacemaker Laboratory and the Syncope Center, where he performs diagnostic studies, interventional therapeutic procedures, consultations, and clinical research. Reiffel is a coauthor of the antiarrhythmic drug section of the American College of Cardiology/North American Society of Pacing and Electrophysiology Clinical Cardiac Electrophysiology Self Assessment Program, and has published numerous articles and abstracts relating to cardiac arrhythmias.

Rickards, Anthony Francis. (*1945; †2004). Dr. Rickards was a leading member of the second generation to enter the field of cardiac pacing. He entered the field of cardiology in the late 1960s, working with Aubrey Leatham and Edgar Sowton in London, UK. In 1978, he devised the earliest practical design for rate modulation of a pacemaker based on sensed physiological changes, specifically changes in the evoked QT interval. Rickards was also a founder of the European Working Group on Cardiac Pacing. He died suddenly and unexpectedly at home on May 28, 2004.

Ritter, Philippe. Medicine doctor (1986), cardiologist (1986), and electrophysiologist. His specializations are pacing, implantable cardioverter-defibrillators (ICDs), and pacing in heart failure. Dr. Ritter is a member of: the French Society of Cardiology; the Nucleus of the French Working Group on Cardiac Pacing; the Nucleus of the European Working Group on Cardiac Pacing; the North American Society of Pacing and Electrophysiology; Mediterranean Society of Pacing and Eletrophysiology; and the International Society of Cardiac Pacing and Electrophysiology. He is also the Chairman of Cardiostim. Ritter has many publications in pacing and ICDs, and in pacing and heart failure.

Rizzon, Paolo. Dr. Rizzon was born in Treviso, Italy, on March 17, 1932. His present position is Professor of Cardiology and Head of the Institute of Cardiovascular Diseases and Director of the Postgraduate Medical School of Cardiology, University of Bari, Italy. He graduated in medicine from Padua University (*magna cum laude*), Italy. He became Associate Professor of Cardiology at the University of Bari in 1970 and was appointed as Director of the Postgraduate Medical School of Cardiology, University of Bari, in 1971. Since 1981 he has been a Professor of Cardiology at the University of Bari. Rizzon was President of the Italian Society of Cardiology (1987–1988), Councilor of the European Society of Cardiology (ESC) (1994–1996), and Vice President of the ESC (1996–1998). His impressive scientific activity includes more than 800 publications, reports, and reviews in national and international journals.

Rosen, Michael R. Dr. Rosen has worked extensively on the analysis of cardiac triggered activity, the influence of the autonomic nervous system on cardiac function, and the developmental biology of the heart. He has long been long involved in pharmacological research and new drug development. Together with Drs. Schwartz and Janse, he is a founder of the "Sicilian Gambit," a think-tank approach to the study of mechanisms of antiarrhythmic function. His current position is Gustavus A. Pfeiffer Professor of Pharmacology and Professor of Pediatrics at the Columbia University College of Physicians and Surgeons in New York City. His seminal contributions to the field include identification of triggered activity as an important mechanism in cardiac arrhythmias, the role of sympathetic innervation in modulation of α-adrenergic receptor-effector coupling in normal and ischemic hearts, and the mechanisms of action of a number of antiarrhythmic drugs. Rosen is currently Editor-in-chief of the *Journal of Cardiovascular Pharmacology* and has been Associate Editor of *Circulation Research*. He has been an author and coauthor on more than 200 peer-reviewed manuscripts. He is recipient of the American Heart Association's Award of Merit and Chairman's Award, and has received the Einthoven Award on the 100th Anniversary of Einthoven's invention of the electrocardiogram (ECG). In 2004, Rosen was awarded with the North American Society of Pacing and Electrophysiology's Distinguished Scientist Award.

Rosenbaum, Mauricio B. (*August 25, 1921, Cordoba, Argentina; †May 4, 2003). In 1951, Dr. Rosenbaum obtained his MD degree from the University of Cordoba, and then trained in internal medicine at the National Hospital in the same city. In 1952, he moved to Buenos Aires, Argentina, where he trained in cardiology at the Ramos Mejia Hospital until 1954. His specific fields of investigation and expertise include epidemiology, diagnostic and therapeutic aspects of Chagasic cardiomy-opathy, electrophysiological and therapeutic effects of antiarrhythmic drugs in life-threatening cardiac arrhythmias, and intermittent atrio-ventricular conduction disturbances. His name is directly linked to the concepts of "hemiblock." Moreover, his latest discovery, the heart's long-lasting memory reflected in the T-wave of the electrocardiogram (ECG), remains a fascinating puzzle in modern electrocardiography (see historical page 192).

Ruskin, Jeremy N. Dr. Ruskin was born in South Africa and earned his MD at Harvard Medical School in 1971. He completed his residency at Harvard before going to USPHS Hospital (Staten Island, New York) as a research associate, where he was introduced to invasive cardiac electro-physiology under the mentorship of Antony N. Damato. After finishing his 2 years at Staten Island, he went back to Harvard (Massachusetts General Hospital) and started a new electrophysiology program, which has grown to be one of the largest and most respected programs in the world. Ruskin is the Director of the Cardiac Arrhythmia Service and Electrophysiology Laboratory at Massachusetts General Hospital, and an Associate Professor of Medicine at Harvard Medical School in Boston. His main research interests, on which he has published intensively, include clinical cardiac electrophysiology, the mechanisms of ventricular arrhythmias, sudden cardiac death in experimental models and humans, and pharmacological and non-pharmacological treatments of arrhyth-mias. Ruskin and his colleagues demonstrated for the first time that monomorphic VT was a cause of unexplained syncope. The fact that many patients who survived cardiac arrest have inducible VT–VF was also first reported by Ruskin and his group. In 2002, he received the North American Society of Pacing and Electrophysiology Pioneer in Cardiac Pacing and Electrophysiology Award.

Ryden, Lars. Dr. Ryden, the Director of Cardiology at the Karolinska Hospital Stockholm, Sweden, has had an extensive career in the investigation of cardiac pacing and associated hemodynamics. He is responsible for spreading the practice of cardiac pacing and coronary arteriography throughout Sweden, from the university hospitals to smaller institutions. He plays an active role in the European Society of Cardiology (past President), in multicenter trials for the management of coronary artery disease, and pacemaker trials for heart failure and the assessment of quality of life following pacemaker implantation.

Saksena, Sanjeev. Dr. Saksena, born September 27, 1952 in India, is a Clinical Professor of Medicine at the University of Medicine and Dentistry New Jersey—Robert Wood Johnson Medical School, in New Brunswick, New Jersey, and Director of the Arrhythmia and Pacemaker Service at the Eastern Heart Institute of the Atlantic Health System, New Jersey. He has pioneered work in both pacing and electrophysiology. One clinical highlight of his career was the implantation of a cardioverter-defibrillator without thoracotomy using a triple electrode system. Saksena has written on all aspects of cardiac electrophysiology. His writings and presentations are particularly noted for their lucidity and the importance of their content. Saksena served as President of the North American Society of Pacing and Electrophysiology from 1997 to 1998.

Sanjeev Saksena

Samet, Philip. Dr. Samet was a long-time Chief of Cardiology at the Mount Sinai Hospital in Miami, Florida. During his tenure he was active in the use of cardiac pacing, as early as 1960, and he fostered the development of the first implanted atrial synchronous pacemaker. His group performed extensive evaluation of the hemodynamics of cardiac pacing and the importance of both change in rate with activity and the contribution of atrial contraction to cardiac output. He also encouraged and participated in the electrophysiological research of bradycardias based on the detection and recording of the His bundle deflection, which, at that time, had been newly introduced. Under his leadership Mount Sinai Hospital became an important pacing and electrophysiology training institution, with many of his trainees now working throughout the USA.

Santini, Massimo. Dr. Santini was born in Rome, Italy, on August 31, 1945. He is currently Director of the Department of Heart Diseases at S. Filippo Neri Hospital, and a Professor of Cardiac Pacing and Electrophysiology at La Sapienza University School of Medicine, both in Rome. He is a past President of the Italian Association of Hospital Cardiologists, and current President of Heart Care, the Italian Heart Foundation. He was Chairman of the Working Group on Cardiac Pacing of the European Society of Cardiology and is a member of the board of the International Society of Cardiac Pacing and Electrophysiology. His main fields of scientific interest are cardiac arrhythmias and pacing, prevention of cardiovascular diseases, and education and research in cardiology. Santini was elected as Secretary General of the Thirteenth World Congress on Cardiac Pacing and Electrophysiology, to be held in Rome 2007.

Massimo Santini

Saoudi, Nadir. The cardiologist Dr. Saoudi was born on September 9, 1952 in Paris, France. His professional address is Centre Hospitalier Princess Grace, 98 000 Monaco (Principauté). Saoudi completed his national military service as an intern in the cardiology department of the General Hospital of Fort de France, Martinique, French West Indies (1978–1980). From 1980 to 1984 he was an intern of the Centre Hospitalier et Universitaire of Rouen in Rouen, France. From 1984 to 1985 he received a Special Research Fellowship in Cardiac Elec-trophysiology (Prof. A. Castellanos) at the Department of Cardiology, University of Miami, School of Medicine, Miami, Florida. From 1985 to 1988 he was Assistant Professor (*chef de clinique*) at the Centre Hospitalier et Universitaire de Rouen in Rouen, France. In 1988, Saoudi became Associate Professor (*practicien hospitalier*) at the Centre Hospitalier et Universitaire de Lyon, in Lyon, France, and then in 1990 he became Associate Professor (*practicien hospitalier*) at the Centre Hospitalier et Universitaire de Rouen In 1993, he was appointed as Professor of Cardiology at the Centre Hospitalier et Universitaire de Rouen, and from 2001 onwards as Chief, Cardiology Department of the Centre Hospitalier Princess Grace of Monaco, in Monaco.

Schaldach, Max Gustav Julius. (*July 19, 1936, Berlin, Germany; †May 5, 2001, vicinity of Nuremberg, Germany). Dr. Schaldach was the son of a Pomeranian couple. He served as Professor and Chairman of Biomedical Engineering at Friedrich–Alexander University, Erlangen–Nuremberg, Professor of Postgraduate Studies at the State School of Medicine, São José do Rio Preto, Brazil, and Professor of Biophysics at Lomonosov University, Moscow. In addition, he was the founder and owner of several high-technology companies, including Biotronik, whose name is associated worldwide with the production of widely used cardiac pacemakers and defibrillators. In 1963, the company developed the first German pacemaker and started marketing the devices, only 5 years after the initial implantation of a pacemaker prototype in Sweden. In its broadest sense, his main focus was electrophysiology. He specialized in electro therapy of the heart, electronic implants, cardiac pacemakers, neurostimulators, defibrillators, artificial organs, circulation-assist devices, and interventional cardiology. These, together with his other areas of research, led to more than 100 patents. Max Schaldach, the physicist and entrepreneur, was killed in a plane crash (see historical page 178).

Scheinman, Melvin M. Dr. Scheinman is a Professor of Medicine and a member of the Clinical Cardiac Electrophysiology Unit of Moffitt–Long Hospital, University of California, San Francisco. He received his medical degree from Albert Einstein College of Medicine in New York, completed his internship and residency at the North Carolina Memorial Hospital, and completed his cardiology fellowship at the University of California, San Francisco. He is currently Associate Editor of the *Journal of the American College of Cardiology*, Editor for electrophysiology of *Pacing and Clinical Electrophysiology: PACE*, and serves on the editorial board of many major English-language cardiology journals. Scheinman is a member of the founding generation of clinical cardiac electrophysiologists and has written on all aspects of cardiac electrophysiology, analysis and management of cardiac arrhythmias, and the use of implantable defibrillators. He introduced the technique of intracavitary ablation of the conduction system and accessory pathways. Scheinman is regarded as a pioneer in the field of ablation and interventional electrophysiology. He reported on the ablation of the atrioventricular (AV) junction using DC shocks for the treatment of refractory supraventricular tachyarrhythmias, first in dogs in 1983, and soon after in humans. Also, he was the first to apply DC shocks transeptally for septal VT and to the coronary sinus for posteroseptal accessory connections, and the first to find a cure for AV node reentry. Scheinman was the first to stress the benefits of pacing combined with β-blockers in patients with the congenital long QT syndrome. He is a past President of the North American Society of Pacing and Electrophysiology (NASPE) (1988–1989) and was honored in 1996 by NASPE as a pioneer in cardiac pacing and electrophysiology (see historical page 199).

Melvin M. Scheinman

Scherlag, Benjamin. A long-time contributor to the field of basic cardiac electrophysiology, Dr. Scherlag introduced the technique of catheter detection of His bundle activity, analysis of Wolff–Parkinson–White tachycardia, and the early demonstration of the role of His bundle deflection in the diagnosis of many arrhythmias. He was involved in demonstrating the significance of the duration of the AV and HV intervals and the prediction of later development of arrhythmia.

Schoenfeld, Mark H. Dr. Schoenfeld was born on December 15, 1953 in New York City. He is currently the Director of the Cardiac Electrophysiology and Pacer Laboratory at the Hospital of Saint Raphael, New Haven, Connecticut, and an Associate Clinical Professor of Medicine at Yale University School of Medicine. He was educated at Yale University and Harvard Medical School and completed his internship, residency, and fellowships in cardiology, cardiac electrophysiology, and cardiac pacing at Massachusetts General Hospital. In 2002–3, Schoenfeld was President of the North American Society of Pacing and Electrophysiology.

Scholz, Hasso. Dr. Scholz was born on August 24, 1937 in Stettin, Germany, the son of a pharmacist. From 1956 to 1966 he dedicated himself to the study of pharmacy and medicine in Heidelberg, Marburg, Berlin, and Mainz, in Germany. In 1975, he spent two research semesters at the Pharmacological Institute of the University of Bern, Switzerland (Prof. H. Reuter). Since 1982 he has been Director of the Institute of Pharmacology at the University Hospital Eppendorf, Hamburg, Germany. His research interests are the action mechanism of pharmaceutical drugs that affect the heart, especially inotropic substances and anti-arrhythmic agents. Furthermore, his attention is currently focused on the regulation of cardiac activity through calcium, cyclic nucleotides, phosphoinositides, and G-proteins. Recently, he has dedicated himself to the myocardial effects of α-sympathomimetics, phosphodiesterase inhibiting agents, and adenosine, as well as the expression of G-proteins and adrenergic receptors under the influence of pharmaceutical substances and with heart failure.

Schwartz, Peter J. Dr. Schwartz was born in the UK and educated in Italy. He is currently Professor and Chair of Cardiology and Director of the School for the Board in Cardiology of the University of Pavia in Italy, and holds a joint appointment as Professor of Physiology and Biophysics at the University of Oklahoma. Schwartz has been involved in the study of the relation between the autonomic nervous system and life-threatening arrhythmias. He has contributed to the study of cardiac reflexes, he developed the concept of sympathetic imbalance, and he has extensively investigated the pathophysiological effects and therapeutic efficacy of left cardiac sympathetic denervation. He also developed the use of baroreflex sensitivity for identifying postmyocardial infarction patients at high risk for sudden death. He and Dr. Arthur J. Moss undertook a joint registry of long QT syndrome, which has been immensely productive in delineating the specific details concerning long QT, its presence in the community, its relationship to sudden death, and the means of its management. The attention given to the long QT syndrome, its management, and, in very large measure, its present recognition, is based on the work of Schwartz and Moss. Schwartz recently provided the first demonstration of differential responses to a variety of interventions according to the specific genetic mutations present in long QT syndrome patients. In 2001, he was honored by the North American Society of Pacing and Electrophysiology with the Distinguished Scientist Award.

Seipel, Ludger. Dr. Seipel was born on May 19, 1939 in Düsseldorf, Germany. Until 2004 he was Chairman of the Department of Internal Medicine III. of the University Hospital, Tübingen, Germany. Seipel and others worked on the ultrasound Doppler cardiogram. His special areas of interest include circulatory diseases in the broadest sense and clinical electrocardiography. His books on *His Bundle Electrography* (1975) and *Clinical Electrophysiology of the Heart* (2nd edn, 1987 in Germany) are well-known. Seipel served as President of the 1998 Conference of the German Society for Cardiology, which focused on cardiac and circulation research.

Sénac, Jean-Baptiste de. (*1693; †1770). Sénac received his medical degree from Montpellier, France. He moved to Paris when he was about

Jean-Baptiste de Sénac

30 years of age. He was an Associate Member of the Academy of Sciences in 1723, and in 1744 moved to Versailles, where he was a Physician at the Royal Hospital of Versailles and was appointed Chief Physician to Louis XV in 1752. His fame in cardiology is based on his *Traité de la Structure du Coeur, de son Action, et de ses Maladies*, a work published in 1749. This book offered a synthesis of what was known about the cardiovascular system in the middle of the eighteenth century.

Senning, Åke (*December 14, 1915, Rattvik, Sweden; †July 21, 2000). Prof. Senning, the son of a physician, was educated in his native Sweden. In 1964, he began the practice of renal transplantation in Zürich, Switzerland, and in 1968 performed the first cardiac transplantation in Europe. In 1958, Senning invented hemodynamic correction of transposition of the great arteries by atrial switch, the "Senning operation." On October 8, 1958, he implanted by thoracotomy the rechargeable cardiac pacemaker that Elmqvist had designed and constructed; this was the first such implant, and was performed on a 43-year-old patient with complete heart block and syncope (Arne Larsson). This pioneer of cardiac surgery and cardiac pacing died several months before his 85th birthday (see historical page 173).

Sethi, Kamal K. Dr. Sethi was born on December 25, 1950 in New Delhi, India. His medical training was completed at the University of New Delhi, India, the Royal Melbourne Hospital, Melbourne, Australia, and St. George's Medical School, London, UK. Sethi's present position is Vice Chairman and Managing Director, Delhi Heart and Lung Institute, New Delhi. He has received relevant scientific awards including the Award of Excellence in Cardiac Pacing, the National Award for Excellence in Medicine, and the Dr. B.C. Roy Award.

Singer, Igor. Dr. Singer was formerly Professor of Medicine, Chief of the Arrhythmia Service and Director of the Electrophysiology and Pacing Program at the University of Louisville, Kentucky. He received his cardiology training at Johns Hopkins Hospital, Baltimore, Maryland. Since January 2003, Singer has taken lead of the Methodist Heart Care program in the newly created role of Executive Medical Director, Methodist Cardiovascular Services, Peoria, Illinois. Singer has authored five widely used textbooks in electrophysiology: *Clinical Manual of Electrophysiology* (Williams & Wilkins, Baltimore, MD); *Implantable Cardioverter Defibrillator* (Futura, Armonk, NY); *Interventional Electrophysiology* (Williams & Wilkins, Baltimore, MD); *Nonpharmacological Therapy of Arrhythmias for the Twenty-First Century:The State of Art* (Futura, Armonk, NY); and *Interventional Electrophysiology* (2nd edn, Lippincott Williams & Wilkins, Philadelphia, PA). In addition, he has published extensively in medical journals and has contributed to numerous highly respected textbooks on cardiac pacing and electrophysiology. Singer was Program Director for the prestigious biennial "International Symposium on Advances in Interventional Electrophysiology" held in Louisville, Kentucky.

Singh, Bramah N. Dr. Singh is a Professor of Medicine and Consultant Cardiologist at the University of California, Los Angeles School of Medicine and at the Veterans Affairs Medical Center of West Los Angeles, where he is also Director of the Cardiovascular Research Laboratory. He earned his PhD in cardiac electropharmacology at the University of Oxford, UK, during which time he formulated the basis for the conventional classification of antiarrhythmic drugs. He is the Editor-in-chief and founding Editor of the *Journal of Cardiovascular Pharmacology and Therapeutics*.

Slama, Robert C. Dr. Slama was born in La Goulette, Tunesia, on July 16, 1929. His present position is Professor Emeritus of Cardiology, University of Paris (VII.). He was Chief of the Department of Cardiology at the Hôpital Lariboisière in Paris, France, from 1979 to 1994. Slama is a fellow of the American College of Cardiology and a fellow of the European Society of Cardiology. He was President of the French Society of Cardiology from 1991 to 1993, and is an officer of the French Legion d'Honneur. Slama was a distinguished guest speaker at the Congress of the American Heart Association in 1989. He has written or cowritten more than 400 scientific articles and eight books, mainly devoted to cardiac arrhythmias. He was the first to describe the Kent bundle and the mechanism of alternating Wenckebach phenomena in the atrioventricular (AV) node.

Sloman, J. Graeme. Dr. Sloman was an early practitioner in cardiac pacing at St. George's Hospital in London, UK. He continued his involvement in pacemaker implantation and follow-up after his return to Australia, and has lectured widely throughout the world, especially in the Asia–Pacific area, where he has been in the forefront of the dissemination of cardiology and pacing information. He is the Editor of *Asia Pace News*.

Smeets, Joseph Léon Robert Marie. Dr. Smeets was born on March 30, 1952 in Heerlen, the Netherlands. He is married and has three children. Smeets completed a Residency in Physiology from 1978 to 1983 at the University of Limburg, the Netherlands (Head Prof. Dr. R.

Reneman). His Residency in Cardiology was completed at the Academic Hospital Maastricht, the Netherlands, from 1983 to 1987 (Head Prof. Dr. H. J. J. Wellens). In 1987, he became a specialist in cardiology; and in 1992 the Director of the Electrophysiology Laboratory of the Cardiology Department of the University Hospital Maastricht, the Netherlands. Smeets was Secretary of the Dutch Working Group on Arrhythmias (part of the Dutch Society of Cardiology) from 1992 to 1997, and from 1997 to 1999 Chairman of the Dutch Working Group on Arrhythmias. He is a reviewer for the leading journals of cardiology, and a member of the European Working Group on Arrhythmias and North American Society of Pacing and Electrophysiology (NASPE)–Heart Rhythm Society. Smeets is author and coauthor in over 140 articles in peer-reviewed journals and has edited two widely used books on Prof. H. J. J. Wellens (*Thirty-Three Years of Cardiology and Arrhythmology*, Kluwer Academic Publishers, 2000; and *H. J. J. Wellens PhD. Thesis and Beyond*, University Press Maastricht, 2000).

Sowton, Edgar. Dr. Sowton of St. George's Hospital, London, UK, was one of the earliest workers in cardiac pacing and contributed analyses of patient survival related to the mode of implanted cardiac pacing. He evaluated early management of ventricular arrhythmias with cardiac pacing and cardiac output as a function of activity and exercise at a fixed rate of stimulation and during atrial synchrony. He coauthored, with Harold Siddons, *Cardiac Pacemakers*, the first book on cardiac pacing.

Surawicz, Borys. Dr Surawicz was born in Moscow, Russia, on November 2, 1917. He received his MD degree at the Stefan Batory University, in Wilno, Poland (today Lithuania), in December 1939. Surawicz immigrated to the USA in 1951. He was a Professor of Medicine at the University of Kentucky College of Medicine from 1966 to 1981, and a Professor of Medicine at the University of Indiana Medical School from 1981 to 1987. Since 1987 he has been a Professor Emeritus and Senior Research Associate at the Krannert Institute of Cardiology, Indianapolis, Indiana. Surawicz is the author of 290 scientific papers, book chapters, and review articles. Furthermore, he is author of two books and coeditor of five books. He has served on editorial boards of 11 scientific journals, received numerous honors and awards, and given many invited lectures. He is an honorary member of the Venezuela Cardiological Association, the New York Cardiovascular Society, the Polish Cardiac Society, and the Nicaragua Cardiological Association. His awards include the Medical Association Scientific Award in 1975, the Osler Award in 1977, and the North American Society of Pacing and Electrophysiology

Distinguished Scientist Award in 1992. Surawicz was President of the Association of University Cardiologists in 1977, and President of the American College of Cardiology from 1979 to 1980.

Sutton, Richard. Dr. Sutton is a Consultant Cardiologist at the Royal Brompton and Harefield National Health Service Trust and the Chelsea and Westminster National Health Service Trust in the UK. He is a fellow of the Royal College of Physicians, London; a Member of the Royal College of Surgeons (MRCS), England; a licentiate of the Royal College of Physicians (LRCP), London; a fellow of the European Society of Cardiology; a fellow of the American College of Cardiology; and a fellow of the American Heart Association (FAHA). Sutton received the Governors' Award of the American College of Cardiology in 1979 and 1982. He is Honorary Consultant Cardiologist at St. Luke's Hospital, London; past Chairman of the Working Group on Cardiac Pacing of the European Society of Cardiology; and has been Editor-in-chief of *Europace* since 1998; previously he was Editor-in-chief of the *European Journal of Cardiac Pacing and Electrophysiology* (1991–1997). Sutton has published scientific articles on cardiac pacing, electrophysiology, syncope, left ventricular function, coronary heart disease, and the effects of drugs on the circulation. His other interests include wine, opera, and travel.

Richard Sutton

Steinbach, Konrad K. Dr. Steinbach was born in Vienna, Austria, on June 1, 1937. He received his education at the University of Vienna. Steinbach became an Associate Professor in 1974. In 1980, he was elected Chairman of the Third Medical Department (Cardiology) of the Wilhelminenspital, Vienna. His main scientific interest is the treatment of cardiac arrhythmias. Steinbach has published more than 350 scientific papers. He is a past President of the Austrian Cardiac Society (1983), and a past President of the International Cardiac Pacing and Electrophysiology Society (1983–1987). He was Chairman of the European Working Group on Cardiac Pacing from 1985 to 1987. Steinbach has received numerous honors and awards. He is an honorary member of the Hungarian, Slovenian, and Slovakian Society of Cardiology.

Steinbeck, Gerhard. Dr. Steinbeck, born in Göttingen, Germany, on August 17, 1946, is currently Head of the Medical Hospital I. of the University of Munich, Klinikum Grosshadern, Germany. He studied medicine at the Universities of Berlin, Hamburg, and Göttingen, and passed his final examination in December 1971. He completed his training at the Physiological Institute of the University of Maastricht, the Netherlands (Profs. Bonke and Allessie) from January 1976 to March 1977, for experimental mapping studies on sinus node function in isolated rabbit heart with multiple microelectrode impalements. After returning to Munich, he undertook experimental and clinical studies in the field of cardiac electrophysiology. His thesis was called "Diagnosis of sinus node function by atrial stimulation: Experimental basis and clinical results." Other projects include programmed ventricular stimulation for diagnosis and treatment of cardiac arrhythmias, and, together with Ralph Haberl, development of a non-invasive method for late potential analysis from the body surface by fast Fourier transform (spectrotemporal mapping). His special interests have included serial drug testing, catheter ablation by radiofrequency current (together with Ellen Hoffmann), and treatment with the automatic implantable defibrillator.

Stokes, William. (*1804, Dublin, Ireland; †1877, Howth, Ireland). Dr. Stokes will be forever associated with two well-known syndromes: Stokes–Adams disease and Cheyne–Stokes respiration. He studied medicine at the Meath Hospital in Dublin, later proceeding to Glasgow, Scotland, and finally to Edinburgh, Scotland, where he obtained his medial degree in 1825. In the fourth volume of the *Dublin Hospital Reports* (1827), Stokes recorded "a patient with permanently slow pulse in which the patient suffered from repeated cerebral attacks of an apoplectic nature, though not followed by paralysis." (See portrait on p. 102.)

William Stokes

Suma, Kozo. Dr. Suma, together with T. Togawa, built an external rechargeable pacemaker in 1960 and later an implantable mercury–zinc powered device. With the commercial availability of implantable pacemakers, local manufacturing ceased. He later became a Professor of Surgery and then Director of Tokyo Women's University. He was President of the International Society of Cardio-Thoracic Surgeons (ISCTS) in 1997. He has published pivotal papers on Sunao Tawara, a pioneering investigator of the conduction system of the heart and the father of modern cardiology. Recently, Suma published an English version of Tawara's famous book *The Conduction System of the Mammalian Heart* (with a foreword by Ludwig Aschoff).

Sykosch, Heinz Joachim. Dr. Sykosch was a young cardiothoracic surgeon when he implanted one of the first cardiac pacemakers in Germany in 1962. In the following year, he and Dr. Fred Zacouto conceived of a pacemaker that stopped stimulating following a spontaneous heart beat and then resumed in the absence of spontaneous beating. In 1968, he demonstrated an entirely metal encased pacemaker, and in 1991 urged that the World Symposium be held in Berlin for its 1999 session. He accomplished that goal and was the honorary President of the World Symposium in Berlin in 1999.

Hilbert J.Th. Thalen

Thalen, Hilbert J.Th. (*April 29, 1939; †October 10, 1982). Dr. Thalen studied medicine at the Medical School of the State University of Groningen in the Netherlands. His 1969 doctoral thesis entitled "The artificial cardiac pacemaker: Its history, development, and clinical application," has become a standard international reference text. In 1973, he organized the Fourth International Symposium on Cardiac Pacing held in Groningen. Thalen served on the editorial boards of the *European Journal of Cardiology*, *Pacing and Clinical Electrophysiology: PACE* and the International Society of Cardiac Pacing. As a long-standing enthusiast of sports medicine, he was a cardiac consultant to the Dutch Society of Sports Medicine.

Touboul, Paul. Dr. Touboul was born on November 27, 1937 in Oran, Algeria. His present position is Professor of Cardiology, University of Lyon (since 1974) and Chief, Cardiovascular Section, Hôpital Cardiovasculaire et Pneumologique Louis Pradel, Lyon, France (since 1979). His areas of special interest are clinical electrophysiology, diagnostics and therapy, and differentiated drug therapy. Touboul is a leader of several important clinical studies, including DAFNE and CAPTIM. He is a member of relevant scientific societies in the field of electrophysiology and pacing in France, Europe, and elsewhere.

Trappe, Hans-Joachim. Dr. Trappe was born on October 17, 1954 in Castrop-Rauxel, Germany. He studied medicine at the University of Göttingen from 1973 to 1979, and was Scientific and Senior Physician at the University Hospital in Hannover, Germany from 1983 to 1996. Trappe completed a research fellowship at the University Hospital Maastricht (Prof. H. J. J. Wellens) from 1986 to 1987. In 1996, he became a Full Professor of Medicine and Head of the Department of Cardiology and Angiology at the University of Bochum, Germany.

Vardas, Panagiotis. Dr. Vardas studied at the Medical School of the University of Athens, Greece, where he received his diploma in 1974. He completed his training at the Cardiology Clinic of Westminster Hospital (1985–1989), London, UK, specializing in electrophysiology and pacing while simultaneously completing his PhD thesis in clinical electrophysiology at the Medical College of Charing Cross and Westminster Hospital. In 1997, he was Chairman of the Congress of the European Working Group of Cardiac Pacing and Arrhythmia (EUROPACE) in Athens. Vardas has published many research articles focusing on pacing, bradyarrhythmias, and atrial fibrillation.

Panagiotis Vardas

Vesalius, Andreas. (*December 31, 1514, Brussels, Belgium; †1564, Sakinthos [island in the Ionian Sea]). Physician, anatomist. From 1528 Vesalius studied classical languages in Leuven, Belgium, and from 1533 on, medicine in Paris. Of particular importance are his work *Tabulae Anatomicae Sex* (Venice, 1538) and, above all, his major pioneering work, *De Humani Corporis Fabrica* (Basel, 1543), which contained more than 250 figures by one or more artists of the Tizian School. The *Fabrica* is recognized as the beginning of modern human anatomy.

Vlay, Stephen Charles. Dr. Vlay was born 1950 in New York City. His present position is Professor of Medicine/Cardiology at University Hospital, SUNY Health Science Center in Stony Brook, New York. He graduated in 1979 and his postdoctoral training took place at New York University, Bellevue Hospital, Medical Center in New York, the Johns Hopkins University School of Medicine, and the Johns Hopkins Hospital in Baltimore, Maryland. Vlay is a fellow and member of the most relevant scientific societies and has editorial responsibilities on the editorial boards of *American Heart Journal* and *Pacing and Clinical Electrophysiology: PACE*. His scientific work includes cardiac arrhythmias and, particularly, quality-of-life issues.

Volta, Alessandro. (*February 18, 1745, Como, Italy; †March 5, 1827, Como, Italy). Count (from 1810), Italian physicist, professor in Como and Pavia. In 1775 Volta invented the electrophorus, which J. C. Wilcke had already indirectly used in 1762. In 1782 he developed from this the capacitor, a very sensitive detection device for weak charges. In 1800 Volta publicized his most important and influential discoveries, the

"voltaic pile" and the "crown of cups." Both devices provided the first forms of high voltage, galvanic battery electricity using a means other than generation through electrostatic machines. The "volt," a unit of electromotive force, is named in his honor.

Waldo, Albert. Dr. Waldo, born in New York, is Walter H. Pritchard Professor of Cardiology, Professor of Medicine, and Professor of Biomedical Engineering at Case Western Reserve University School of Medicine, Cleveland, Ohio. He is also the Director of the Clinical Cardiac Electrophysiology Program and is a staff physician of the University Hospitals of Cleveland. Waldo received his MD from the State University of New York College of Medicine, and completed postdoctoral fellowships in cardiology and cardiac electrophysiology at Columbia-Presbyterian Medical Center in New York City. He is a founding member and past President of the North American Society of Pacing and Electrophysiology and has served as President of the Cardiac Electrophysiology Society. Among his most notable achievements are the recognition of the complexity of atrial activation that produced unexpected P-wave morphologies on the electrocardiogram (ECG), the intraoperative mapping of the His bundle in complex congenital heart disease, and the discovery of the low-level "fractionated" potentials generated by ischemic myocardium. Waldo characterized and classified atrial flutter in humans. Furthermore, he described the characteristic pause antecedence of torsade de pointes and the introduction of the concept and practice of entrainment of cardiac arrhythmias, which has been

Albert Waldo

a landmark in the century-long study of the process of reentry and an invaluable aid to clinical electrophysiologists.

Waller, Augustus Desiré. (*1856, Paris, France; †1922). Dr. Waller was the son of the British physiologist Augustus Volney Waller, discoverer of Wallerian degeneration of nerves, who was studying in Paris at the time of his son's birth. A. D. Waller qualified in medicine at Aberdeen in Scotland, and acquired his interest in cardiac physiology from John Burdon Sanderson at University College London, UK. His postgraduate studies were performed under Carl Ludwig in Leipzig, Germany. Waller became a lecturer in physiology at St. Mary's Hospital, London, in 1885. In 1887 he recorded the first human electrocardiogram (ECG). He called this tracing an "electrocardiogram." Waller published his discovery in the *The Journal of Physiology* in 1887. He would often use his pet bulldog, Jimmie, as the subject when he demonstrated his method at lectures. Waller wrote *Introduction to Human Physiology* (1891) and *Animal Electricity* (1896). He used the Lippmann capillary electrometer to record cardiac potentials from the exposed hearts of animals. He is buried in Finchley Cemetery in London (see historical page 146).

Watanabe, Yoshio. A distinguished electrocardiographer, Dr. Watanabe has been Professor of Medicine at Hahnemann Medical College and Professor of Medicine and Director of the Cardiovascular Institute at Fujita Health University. He and Leonard Dreifus have conducted courses in electrocardiogram (ECG) interpretation for many years.

Weber, Arthur. (*August 3, 1879, Fechenheim, Germany; †June 7, 1975, Eschwege, Germany). Prof. Dr. Weber was the leading German specialist in electrocardiography during the 1920s and 1930s. He received his medical education in Marburg, Leipzig, and Greifswald, and received his Doctor of Medicine in 1904. He lived and worked for almost 60 years in Bad Nauheim, the famous German spa that is recommended for cardiovascular diseases. Weber became a Full Professor of Balneology at the University of Giessen in 1943. He must be regarded as a founder of the first independent Department of Cardiology in Germany. His clinical and scientific interests included phonocardiography, venous pulse evaluation, and, primarily, electrocardiography. Weber was cofounder (with Bruno Kisch) of the German Society of Cardiology in 1927, in Bad Nauheim. He published more than 150 scientific articles and books and received numerous honors and awards. A prestigious, non-industry-sponsored award of the German Society of Cardiology was named in his honor.

Wellens, Henrick Joan Joost. Dr. Wellens, born November 13, 1935 in te's-Gravenhaage, is a Professor of Cardiology in Maastricht, the Netherlands. He described the mechanism of reentry tachycardia, including Wolff–Parkinson–White syndrome, and is a founder of clinical electrophysiology. That analysis began the modern era of tachycardia management. He has continued with analysis and management of ventricular and atrial arrhythmias and is a master electrocardiographer who, with Mark Josephson, has taught an annual course for the past 20 years. Currently, he is extensively involved in the management of atrial fibrillation. With the publication of his paper on the new "atrioverter," Dr. Wellens presented pioneering work in the important new field of clinical cardiac electrophysiology. This could not have been achieved without his previous fundamental investigations on arrhythmic mechanisms and effects of therapeutic interventions in heart disease. As a result, it is appropriate that Wellens was awarded—among many other honors and prizes—the titles of Pioneer in Cardiac Pacing and Electrophysiology (1995) and Distinguished Teacher (2000) by the North American Society of Pacing and Electrophysiology. Recently, he was knighted by the Queen of the Netherlands. Wellens has indeed provided a higher standard of excellence in the world of arrhythmology (see historical page 166).

H.J.J. Wellens

Wenckebach, Karel Frederik. (*March 24, 1864, the Hague, the Netherlands; †November 11, 1940, Vienna, Austria). A physician, famous for his first description of premature heartbeats in the last decade of the nineteenth century, Dr. Wenckebach studied physiology under T. W. Engelmann at the University of Utrecht, the Netherlands. From 1901 to 1911 he was a Professor of Medicine at the University of Groningen, Austria; and from 1911 to 1914 a Professor of Medicine at the University of Strasbourg, Austria. In 1914, he became Director of the I. Department of Medicine at the University of Vienna, where he lived and worked for the remainder of his life. On May 14, 1898, Wenckebach recorded a regular intermittent pulse in a 40-year-old woman who had a nervous disposition. His careful description of the radial pulse recording revealed what is now known as "Wenckebach block" or "Wenckebach periods." Under his leadership for those years, the I. Department of Medicine became famous worldwide. Wenckebach revolutionized our understanding of cardiology not only with his first descriptions of premature beats, but also with his standard work on *Die unregelmässige Herztätigkeit und ihre klinische Bedeutung*, which was published for the first time in 1914, and revised and re-published in 1927. In 1918, together with W. Frey, Wenckebach introduced quinidine as the treatment of choice for atrial fibrillation and other arrhythmias. He may be considered a founder of modern clinical cardiology and arrhythmology (see historical page 152).

Wilber, David J. Dr. Wilber was born 1951 in Wisconsin, USA. His current position is George M. Eisenberg Professor of Cardiovascular Science, Director, Division of Cardiology and Cardiovascular Institute, Loyola University Medical Center, Chicago, Illinois. His medical training started at the Northwestern University Medical School, Chicago, followed by a residency at the Northwestern Memorial Hospital, Chicago, and a cardiology fellowship at the University of Michigan. Wilber got his training in clinical cardiac electrophysiology at the Massachusetts General Hospital in Boston, Massachusetts. He received remarkable honors and awards: Named to Best Doctors in America 1992–2004 and America's Top Doctors 2000–2004. Wilber is a member of the Association of University Cardiologists, and a fellow of the American College of Cardiology and of the American Heart Association. His research interests include mechanisms and ablation of ventricular tachycardia, risk stratification and primary prevention of sudden death, mapping techniques, and mechanisms and ablation of atrial fibrillation.

Wilkoff, Bruce L. Dr. Wilkoff was born on December 12, 1954. His current position is Director of Cardiac Pacing and Tachyarrhythmia

Devices and Medical Director of Clinical Electrophysiology Research; Director, Cardiovascular Computer Unit, and from March 1992–present Associate Professor of Internal Medicine, Ohio State University/Cleveland Clinic Foundation Health Science Center, Cleveland, Ohio. Wilkoff holds four US patents on implantable medical devices, non-programmable automated heart rhythm classifiers, implantable anti-tachycardia therapy devices, and an energy efficient multiple sensor cardiac pacemaker. He is reviewer of the leading journals of cardiology in the USA and Europe. He serves in the editorial board of *Heartweb Electronic Journal*, *The Journal of Cardiovascular Electrophysiology*, and *Heart Rhythm*. Wilkoff's research activities concern electrotherapy of cardiac arrhythmias by including antitachycardia pacing and implantable cardioverter-defibrillators.

Wit, Andrew L. Dr. Wit was born in Oceanside, New York, on January 18, 1942. He was a cardiac electrophysiology trainee under Anthony N. Damato, MD from 1968 to 1970. His current position is Professor of Pharmacology and Associate Chairman, Department of Pharmacology, College of Physicians and Surgeons at Columbia University in New York. He is a fellow of the American Heart Association and a member of the Council on Basic Cardiovascular Sciences of the American Heart Association. Wit has received numerous awards, including the 1998 Gender Equity Award from the American Medical Women's Association, College of Physicians and Surgeons of Columbia University, the Presidential Award for Outstanding Teaching from Columbia University (1997), the Charles W. Bohmfalk Award for Distinguished Contributions to Teaching in the Preclinical Sciences, College of Physicians and Surgeons of Columbia University (1994), the American College of Cardiology Honorary Fellowship Award (1993), and the Thomas Lewis Award of the Los Angeles Cardiac Electrophysiology Society (1992). Wit is a world renowned investigator. He has published pioneering work in the leading peer-reviewed English-language journals of cardiovascular sciences.

Withering, William. (*1741; †1799). A physician to the General Hospital at Birmingham, UK, Dr. Withering first described the effects of digitalis purpurea on the heart in 1785 in his well-known book, *An Account of the Foxglove and Some of Its Medical Uses: With Practical Remarks on Dropsy and Other Diseases*. In this book, Withering reported on a decade of observations and experience with digitalis purpurea on numerous dropsy patients. Withering's brother was Director of the Botanic Garden in Birmingham. It was Withering who initiated digitalis therapy. (See plaque on p. 111.)

Witte, Joachim. Dr. Witte (now retired) was the leading pacemaker practitioner in the former German Democratic Republic until 1989; during this time, he implanted more than 9000 pacemakers. His special interest was in follow-ups using the oscilloscopic display of the pacemaker stimulus and in the development of a large database encompassing all of the pacemaker implants in his country, allowing follow-up for reliability and evolution of disease. He was also extensively involved with atrial pacing and demonstrated the long-term stability and reliability of atrial pacing leads and the long-term stability of atrioventricular (AV) conduction where atrial pacing is used for management of sinus node dysfunction. He was an early advocate of endocardial atrial synchronous pacing and published widely on this subject.

Wyse, D. George. Dr. Wyse is a Professor of Pharmacology and Medicine (Cardiology) in the Departments of Medicine (Cardiology)/ Pharmacology and Therapeutics at the University of Calgary in Calgary, Alberta, Canada. He obtained his PhD degree in pharmacology from McGill University and participated in postdoctoral research at the University of New Mexico before entering medical school. In 1992, he took a sabbatical leave from the university to pursue research at the Clinical Trials Branch of the National Heart, Lung, and Blood Institute, where he was principally involved in the planning and completion of the cardiac arrhythmia suppression trial (CAST) and the antiarrhythmics versus implantable defibrillators (AVID) trial. In addition, he chaired the Planning and Steering Committees for the Atrial Fibrillation Follow-up Investigation of Rhythm Management (AFFIRM) trial. Wyse is on the editorial boards of *The Journal of Cardiovascular Electrophysiology* and the *North American Society of Pacing and Electrophysiology Tapes*, and is Associate Editor of *Journal of Interventional Cardiac Electrophysiology*.

Yee, Raymond. Dr. Yee was born on October 29, 1953 in Edmonton, Alberta, Canada. His professional address is Department of Medicine, Division of Cardiology, London Health Sciences Center, University Campus, 339 Windermere Road, London, Ontario, Canada. Yee received his education at the University of Alberta, Edmonton (1973–1977) where he received his MD degree with distinction. His academic training includes the position as Resident in Internal Medicine and Cardiology, University of Western Ontario and London Health Sciences Center, London, Ontario (1979–1983) and as Research Associate, Division of Cardiology, Duke University Medical Center, Durham, North Carolina (1983–1986). In 1996, he became Director of the Arrhythmia Service and in 1997 Professor of the Department of Medicine, University of Western Ontario, London, Ontario. In 1985, he became a Fellow of the Royal College of Physicians and Surgeons of Canada (Cardiology) (FRCPC) and in 1988 a Fellow of the American College of Cardiology (FACC).

Zacouto, Fred Isaac. Dr. Zacouto, who was born in 1924 in Berlin, Germany, is a physician and an electrical engineer who developed a cardiac pacemaker in an attempt to resuscitate the hearts of stillborn fetuses in 1952. Later that year, he conceived of and developed a device to pace externally following a period of asystole, and attempted it in an internal version in 1959. He later added an automatic defibrillation system based on the detection of a rate greater than 180 beats per minute. Another accomplishment was orthorhythmic pacing, an antitachycardia mode widely evaluated before the era of implantable defibrillation.

Zipes, Douglas Peter. Born February 27, 1939, Dr. Zipes is a premier electrophysiologist, researcher, teacher, author, and Editor of *The Journal of Cardiovascular Electrophysiology. Pictured below far right.* He introduced the concept of low-power defibrillation for ventricular arrhythmias and

has been noted for leadership of multicenter, prospective, randomized clinical trials, such as the antiarrhythmics versus implantable defibrillators (AVID) trial. He was Chairman of the American Board of Internal Medicine Committee, which developed the examination in clinical cardiac electrophysiology, and has been President of the North American Society of Pacing and Electrophysiology. Zipes became President of the American College of Cardiology in March 2001 (Orlando, Florida) (see historical page 213).

Zoll, Paul M. Dr. Zoll was born and educated in Boston, Massachusetts. He attended Harvard Medical School and trained at the Boston Beth Israel Hospital, where he remained for the duration of his career. In an epochal publication in 1952, he described cardiac resuscitation via electrodes on the bare chest with 2-ms-duration pulses of 100–150 volts across the chest, at some 60 stimuli per minute. This initial clinical description launched widespread evaluation of pacing and the recognition by the medical profession and the public that the asystolic heart could be stimulated to beat, and was the basis for many developments to come. In 1956, Zoll published a description of a transcutaneous approach that terminated ventricular fibrillation with a much larger shock. He later described similar termination of ventricular tachycardia. He was awarded the Lasker Award in 1973, and was recognized as a Pioneer in Cardiac Pacing by the North American Society of Pacing and Electrophysiology in 1989 (see historical page 171).

10TH ANNUAL EP FELLOWS SEMINAR • MASTERS SESSION • NOVEMBER 8, 1997

The following signatures are illustrated: 1. **Masood Akhtar**. Studies on physiology and reentry in the bundle branch system. 2. **Maurits A. Allessie**. Studies on atrial flutter. 3. **Agustin Castellanos**. Studies on parasystole. 4. **Philippe Coumel**. Arrhythmias on long-term electrocardiogram (ECG) monitoring. 5. **Jim L. Cox**. Localization and surgical excision of accessory pathways in the AV groove. 6. **Guy Fontaine**. Electrical fulguration of ventricular tachycardia. 7. **John J. Gallagher**. Electrical ablation of the bundle of His. 8. **Gerald Guiraudon**. Surgical treatment of ventricular tachycardia by encircling ventriculotomy. 9. **Michiel J. Janse**. Studies on ischemic arrhythmias in the pig heart. 10. **Mark E. Josephson**. Resection of the site of origin (of ventricular tachycardia). 11. **Richard Langendorf**. Reentry. 12. **William J. McKenna**. Arrhythmias and sudden death in hypertrophic cardiomyopathy. 13. **Michel Mirowski**. The implantable defibrillator. 14. **Robert J. Myerburg**. Studies on arrhythmias in the dog's heart. 15. **Peter J. Schwartz**. The long AT syndrome and torsade de pointes. 16. **Michael B. Simson**. Late potentials on the surface electrocardiogram (ECG). 17. **Harold C. Strauss**. Sinus node function. 18. **Albert L. Waldo**. Entrainment (continuous resetting of a reentrant circuit). 19. **Henrick J. J. Wellens**. Electrophysiologic studies in humans with multielectrode catheters. 20. **Andrew L. Wit**. Anisotropy of heart muscle in infarcted regions. 21. **Douglas P. Zipes**. The implantable cardioverter. From Brugada, P. (1986) How some arrhythmologists should sign and why. *American Journal of Cardiology*, **58**, 381–2, with permission.

Part 2

Dictionary of Electrophysiology and Pacing

Aberrancy Abnormal *conduction* in the native tissues resulting in a change in the sequence of activation (and hence, *morphology*) of the atrial or *ventricular depolarization*. This is due to functional or pathological increases in the physiologic *refractory period* of the portion of the conduction system.

Ablation Removal or destruction of tissue.

Accessory pathway A *conduction* pathway connecting the atrium and ventricle that totally or partially bypasses the *AV node*. Conduction through accessory pathways can interfere with normal conduction and cause preexcitation of myocardial tissue. Accessory pathways may provide *anterograde* or *retrograde conduction* of impulses and frequently are the anatomical basis for reentrant arrhythmias.

Action potential The changes in electrical potential generated by the muscle cell membrane or nerve cell tissue in response to *intrinsic* or extrinsic stimulation. There are five phases: phase 0 is the period of rapid *depolarization* (polarity changes from negative to positive) and phases 1–4 return the cell to resting membrane potential.

AH interval The lengths of time taken for a *depolarization* wave to travel from the lower right atrium at the interatrial septum to the *His bundle*, as measured on an *intracardiac electrogram*. This *interval* approximates AV nodal *conduction* time and is normally 50–120 ms.

Annulus fibrosis Non-conductive fibrous or fibrocartiliginous tissue which separates the atria from the ventricles. Also known as the "skeletal structure" of the heart.

Antitachycardia pacemaker (ATP) is an option for a small number of patients with drug-refractory tachycardia. The pacemaker may be temporary or permanent, in conjunction with an implantable cardioverter-defibrillator (ICD).

Antitachycardia pacing (ATP) Therapeutic intervention using standard *bradycardia* pacing *algorithms* and energy levels in an effort to bring the heart out of a *tachycardia* and restore its normal rhythm. ATP is a form of implantable cardioverter-defibrillator (ICD) therapy that uses pacing to terminate a ventricular arrhythmia. ATP is different than a shock; most of the time, ATP is painless.

Arrhythmias can be subdivided into disorders involving *impulse formation* (increased automaticity, abnormal automaticity, triggered activity) and *conduction disorders* (circus movement, reentry).

Ashman phenomenon The Ashman phenomenon describes a wide, aberrantly conducted supraventricular beat occurring after a QRS complex that is preceded by a long pause. The bundle branches reset their rate of repolarization according to the length of the preceding beat. If, before repolarization is complete, another supraventricular impulse should pass through the AV node, conduction will be blocked along the normal pathways and a wide, bizarre QRS complex will be inscribed.

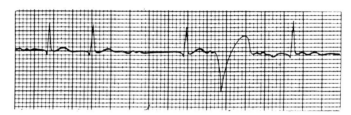

Ashman phenomenon.

Atrial contribution The augmented filling of the ventricles caused by the contraction of the atria during the final phase of ventricular diastole and immediately prior to ventricular contraction. The atrial contribution may add 20% or more to *cardiac output*. It is sometimes called the atrial kick.

Atrial demand (AAI) pacemaker provides the same type of pacing support in the atrium with atrial pacing and sending (AA) and inhibition (I) in the case of intrinsic atrial events. This type of pacemaker has the advantage of preserving the atrial-ventricular sequence of contraction, which, in turn, delivers improved hemodynamics. The prerequisite for this device is unimpaired atrioventricular conduction.

Automated external defibrillator (AED) An external device that can be used by minimally trained people in emergency situations to delivery an electric shock to "reset" a heart that is fibrillating (quivering instead of pumping).

Automaticity Physiology: the ability of tissue to spontaneously generate an electrical impulse and propagate an *action potential*. Automaticity is an inherent property of the normal *sinoatrial (SA) node*.

Autonomic nervous system The nervous system regulating tissues and functions not normally under conscious control, such as heartbeat and blood pressure. It is divided into the sympathetic and parasympathetic nervous systems, which have opposite effects on the cardiovascular system. The sympathetic nerves, when stimulated, increase *heart rate* and myocardial contractility, constrict blood vessels, and raise blood pressure. The parasympathetic nerves decrease heart rate, relax blood vessels, and lower blood pressure.

AV (atrioventricular) block A partial or total interruption of the *conduction* of electrical impulses from the atria to the ventricles, which is defined in degrees: first degree, second degree, *Mobitz I, Mobitz II*, or third degree.

AV dissociation A conduction in which the atria and ventricles beat independently of each other without any synchronization of rhythms.

AV nodal reentry tachycardia (AVNRT) A type of rapid atrial *arrhythmia* characterized by brief periods of *sudden onset* atrial *tachycardia* alternating with periods of *normal sinus rhythm*. The sudden onset of the tachycardia is caused by *reentry* within the *AV node*, or between the AV node and a bypass tract. Sometimes called *paroxysmal* atrial tachycardia, AV *reentry tachycardia*, or *paroxysmal supraventricular tachycardia*.

AV node A collection of specialized cardiac cells located in the right atrium between the *coronary sinus* and the tricuspid valve's septal cusp, which forms a portion of the heart's electrical *conduction* system. In conducting the electrical impulse from the atria to the ventricles, the *AV node* introduces a delay, which allows the atria to contract first, thus augmenting the filling of the relaxed ventricles. It also acts as a physiologic gate to prevent the ventricles from contracting too rapidly in response to pathological atrial tachyarrhythmias.

AV sequential dual-chamber pacemaker (DDD), sometimes called the physiological pacemaker, combines all the pacing states of a demand pacemaker (D = double chambers paced, D = double chambers sensed, D = double modes of response). Demand pacing is delivered to the right atrium and right ventricle. Intrinsic cardiac activity in either the atrium or the ventricle may inhibit the delivery of an output pulse. Furthermore, atrial tracking allows the ventricle to be paced appropriately in response to intrinsic atrial activity.

AV synchrony The condition of the healthy heart in which the atria and ventricles beat in such a way that their individual depolarizations are coordinated to allow atrial contraction to augment the filling of the ventricles.

Backup pacing An independent secondary pacing system that provides protection in the event of component malfunction or *electromagnetic interference*.

Battery The internal *power* source of implantable devices. Most pacemakers today use single-cell, lithium–iodine batteries, while implantable cardioverter-defibrillators (ICDs) use a lithium vanadium pentoxide battery.

Bigeminy Cardiac rhythm characterized by pairs of complexes. This commonly refers to a ventricular ectopic beat coupled to a sinus beat, but it could also be an atrial ectopic beat or paced ventricular beat coupled to either a conducted ventricular complex or a ventricular ectopic beat.

Biphasic A waveform *morphology* having both a positive and negative deflection.

Bipolar Having two poles. In pacing and ICDs a bipolar *lead* contains two *distal electrodes*. Bipolar pulse generators are devices that can accommodate such a lead.

b.p.m. Abbreviation for beats per minute. Usually refers to an intrinsic *heart rate*, while pulses per minute (p.p.m.) usually refers to the paced rate.

Bradycardia Slow *heart rate*, usually defined as less than 60 beats per minute or any rate that is too slow to be physiologically appropriate for the patient's age, condition, and activity level. It results from a dysfunction of impulse formation (sinus node disorders) or impaired conduction. Sick sinus syndrome, sinoatrial block, and atrioventricular (AV) block are of special clinical significance.

Bundle branch block (BBB) An intraventricular *conduction* disorder in which the conduction of electrical impulses through the right or *left bundle branch* is partially or completely interrupted. Bundle branch block causes one of the ventricles to contract before the other.

Burst pacing Several sequential, rapid stimuli delivered to the heart by an external or internal device in an effort to terminate or initiate a tachycardia.

Burst rate The rate of the *pacing train* delivery by an antitachycardia pacemaker or a defibrillator during an attempt to terminate a sensed *tachycardia*.

Cardiac arrest The sudden cessation of all ventricular activity. This could be *asystole* (as with *complete heart block*) but is most usually *ventricular fibrillation*. Cardiac arrest due to ventricular fibrillation is amenable to *defibrillation* by an external or implantable device. If circulation is not supported by external cardiac massage or cardiopulmonary resuscitation (CPR), or if a normal rhythm is not restored within minutes, the result will be death.

Cardiac conduction system The route of normal flow for electrical impulses through the heart. It includes the *sinoatrial node*, intranodal atrial *conduction* tracts, interatrial conduction tract (Bachmann's bundle), *AV node, His bundle*, right and left bundle branches, and the *Purkinje* network.

Cardiac cycle One complete heartbeat, seen on the electrocardiogram (ECG) as a *P-wave*, a *QRS complex*, and a T-wave. In a normal cardiac cycle, the atria depolarize and contract, delivering blood into the relaxed ventricles (P-wave); the atria repolarize and relax while the ventricles, now filled to capacity with blood, contract and pump blood into the systemic and pulmonary circulation systems (QRS complex). After the ventricular contraction, the ventricles repolarize and relax (T-wave). The cycle begins again at the end of ventricular *repolarization* with the onset of passive ventricular filling.

Cardiac index A measurement of the *cardiac output* adjusted for the size of the individual. This is calculated by dividing the cardiac output by the body surface area and is reported as liters per minute per meters-squared (L/min/m²).

Cardiac mapping An invasive electrophysiological procedure used to identify accessory pathways in *Wolff–Parkinson–White syndrome* and other dysrhythmias, to identify atrial activation, to delineate the course of AV *conduction*, to assess the sequence of VA activation, and to locate the origin of tachycardias. Cardiac mapping involves the placement of numerous electrodes on different sections of the endocardium, the delivery of electrical stimuli through the electrodes, and the subsequent analysis of conducted stimuli.

Cardiac output (CO) The volume of blood, measured in liters, ejected by the heart per minute. Cardiac output is determined by multiplying the *heart rate* and the *stroke volume*.

Cardioversion The delivery of an electrical shock to the heart to synchronize the action potentials of all myocardial cells and to terminate arrhythmias. Conversion of *ventricular tachycardia* is often referred to as cardioversion, while conversion of *ventricular fibrillation* is often referred to as *defibrillation*. Synchronization of the shock helps to prevent ventricular fibrillation. In ICDs, the term cardioversion is used to differentiate a therapy for tachyarrhythmias from those for fibrillation.

Cardioverter-defibrillator A device for treating patients at risk of sudden death from dangerous ventricular arrhythmias. The device is capable of delivering an electrical shock to the heart in order to return it to normal rhythm.

Carotid sinus syndrome A hyper-reflex of the pressure receptors in the carotid sinus. It appears on the electrocardiogram as an asystole during transient sinus arrest or third-degree sinoatrial block; it may also appear as a transient atrioventricular (AV) block. Clinically, it means reduced cerebral blood flow, the effect of which may procedure symptoms ranging from mild sensations of dizziness to severe syncopal attacks.

Catheter ablation is a non-invasive procedure using catheterization for treatment of symptomatic, refractory supraventricular tachycardias. The procedure is particularly useful for preexcitation syndrome, in that it interrupts the accessory pathway with high-frequency current and thus eliminates the tachycardia. Up until now, catheter ablation has mainly been used in special cases of ventricular tachyarrhythmias.

Chronaxie The threshold *pulse width* on the *strength–duration curve* at twice the *rheobase* value (in volts).

Congestive heart failure (CHF) The failure of the heart to maintain an adequate *output*, resulting in diminished blood flow to the tissues and congestion in the pulmonary and/or the systemic circulation.

Crista terminalis An anatomical site located in the right atrium at the juncture of the smooth wall tissue and the pectinate muscle of the right atrial appendage.

Defibrillation The termination of ventricular and sometimes atrial *fibrillation* using a *high-voltage* electrical shock. This shock can be delivered to the heart through chest-wall electrodes, directly to the heart during open-heart surgery, or by an automatic implantable cardioverter-defibrillator (ICD).

Depolarization The sudden change in electrical potential from negative to slightly positive which occurs during phase 0 of the *action potential*. In the heart, electrical depolarization initiates the mechanical contraction. Waves of depolarization spread from cell to cell. When this occurs in the atria, a native *P-wave* is seen on the electrocardiogram (ECG); when this occurs in the ventricles, this is seen as a native *QRS complex* on the ECG. Depolarization can also be initiated by a *pacing stimulus*.

Direct current (DC) An electric *current* flowing in one direction only and substantially constant in value. Sometimes used more generally to denote monophasic current or voltage pulses whose value changes with time but whose *polarity* does not.

Dual-chamber pacing A pacing modality capable of delivering pacing stimuli to two chambers, usually the right atrium and ventricle.

Ectopic focus Source of a stimulating impulse for the heart other than the *sinoatrial node*. Typical ectopic foci are the atria, the atrioventricular (AV) junction, and the ventricles, each of which has a characteristic, inherent rate. The rhythm originating from an ectopic focus is either an escape rhythm or an accelerated *ectopic rhythm*.

Ectopic rhythms An abnormal rhythm arising from a site other than the sinoatrial node and, normally, a site other than the *atrioventricular (AV) node* as well. Capable of initiating a spontaneous sustained rhythm in either the atria or the ventricles.

Ejection fraction The percentage of blood volume ejected from the ventricle after contraction. It is calculated by dividing the *stroke volume* by the end diastolic volume. A normal ejection fraction is greater than 55%.

Electrocardiogram (ECG or EKG) A graphic depiction of the electrical signal emitted by active cardiac tissue and recorded through electrodes placed on the body surface. It is also known as a surface electrocardiogram.

Electrode (a) The portion of the lead at or near the tip through which the electrical impulse is conducted and/or *intrinsic* signals are measured. *Unipolar* leads have one electrode, while *bipolar* leads have two electrodes (the *distal* tip electrode and the *proximal* ring electrode) which are physically located on or in the heart. Electrodes may be made of metal, carbon, or other material. (b) Any device used to transmit electrical *energy* to the heart. (c) Any device used to record electrical signals from the heart, such as surface electrocardiogram (ECG) electrodes taped to the skin.

Electrophysiologic study (EPS) An invasive study of the electrical properties of the heart to determine specific functional characteristics that include *automaticity,* myocardial *excitability, refractoriness,* and conductivity. A variable number of pacing catheters are placed in the heart for stimulation and recording. Electrophysiologic studies can assess the mechanism of a *tachycardia* by analyzing its induction, termination sequence and pattern of *conduction,* and help guide the selection of appropriate antiarrhythmic therapy.

Entrainment Continuous resetting of a reentrant circuit by a pacing drive faster than the *tachycardia cycle length.*

Excitability The property of myocardial and other electrically active tissues which allows them to be stimulated to depolarize by either intrinsic *conduction* or an artificial stimulus.

External pacemaker A non-implantable pacemaker used temporarily outside the body to stimulate the heart and sense *intrinsic* cardiac activity. It is also known as a temporary pacemaker.

Fibrillation A type of cardiac *arrhythmia* characterized by rapid, unsynchronized quivering of atria or ventricles. Atrial fibrillation may be asymptomatic, but *ventricular fibrillation* is typically fatal if not corrected within minutes.

Flutter Rapid but regular rhythm (250–350 beats per minute in the atria or 200–300 beats per minute in the ventricles) often seen on the electrocardiogram (ECG) as a saw-toothed pattern. Ventricular flutter may result in death unless corrected within minutes.

Heart block A disruption in the *propagation* of electrical impulses through the atrioventricular (AV) *conduction* system, which can be classified in various grades from first degree to third degree.

Heart failure A condition of decreased *cardiac output* with respect to the metabolic needs of the body.

Heart rate The number of times the heart beats in a minute, normally expressed in the USA as beats per minute or *b.p.m.* and in European countries as min^{-1}.

His bundle Located just distal to the *atrioventricular (AV) node,* this bundle of specialized tissue provides the electrical connection between the atria and the ventricles. The electrocardiogram (ECG) deflection of the His bundle is used as a marker to distinguish normal from abnormal *conduction*. It is also called the *bundle of His*.

His bundle electrogram (HBE) The *intracardiac electrogram* of the *His bundle*, recorded using an endocardial *bipolar* or multipolar *catheter* or with specially placed, filtered, and processed surface electrocardiogram (ECG) leads. HBEs are used in the assessment of wide complex rhythms, *atrioventricular (AV) conduction*, and *preexcitation* syndromes. They allow evaluation of AV nodal conduction time, the *AH interval*, and the *HV interval*.

Holiday heart syndrome Acute arrhythmias or conduction block in persons without organic heart disease occurring in connection with excessive alcohol consumption and disappearance with a sensation of alcohol effect.

Holter monitoring *Noninvasive* ambulatory electrocardiographic monitoring to assess rhythm disorders, usually of more than 24 hours. During normal patient activity or selected diagnostic procedures, intermittent and transient arrhythmias are recorded and correlated with clinical symptoms reported by the patient.

HV interval The time required for a wave of *depolarization* to travel from the *atrioventricular (AV) node* to the *His bundle*, measured as the interval from the low atrial deflection to the His bundle deflection on a *His bundle electrogram*.

Hybrid therapy (*Hybrida, hybridae* (latin): mixture, bastard [two origins]). Hybrid therapies in atrial fibrillation are: (i) linear lesions and antiarrhythmic drugs; (ii) antibradycardia pacing and antiarrhythmic drugs; (iii) pharmacological and ablative therapy. For example, prevention of recurrence with amiodarone/Class IC drug-induced atrial flutter and ablation (isthmus ablation followed by flutter inducing drug therapy; efficacy concerning atrial fibrillation suppression ca. 80%).

Implantable cardioverter-defibrillator (ICD) An ICD is an electrical device which automatically delivers an electrical shock when it detects a dangerous ventricular arrhythmia. The shock is usually sufficient to terminate the ventricular tachyarrhythmia. Modern ICD systems combine bradycardia pacing, antitachycardia pacing, and defibrillatory functions in one single, implantable device. ICDs are 99% effective in stopping life-threatening arrhythmias, and are the most successful therapy to treat ventricular fibrillation, the major cause of sudden cardiac death. ICDs continuously monitor the heart rhythm, automatically function as pacemakers for heart rates that are too slow, and deliver life-saving shocks if a dangerously fast heart rhythm is detected.

Intracardiac electrogram (IEGM) The graphic depiction of the electrical signal emitted by active cardiac tissue and recorded through electrodes placed on or within the heart. EGM counts are used as a measure of sense amplitude in pacemakers and implantable cardioverter-defibrillators (ICDs). Sometimes called *electrogram*, egram, or EGM.

Jervell and Lange-Nielsen syndrome A recessive genetic congenital syndrome characterized by QT prolongation and deafness. The prolonged QT syndrome predisposes to *bradycardia*-dependent ventricular tachyarrhythmias including *polymorphic VT* (torsade de pointes) and *ventricular fibrillation*. A consequence of this syndrome is sudden death at a young age. This syndrome is a subset of *long QT syndrome.*

Kent bundle or Kent fiber Electrically conductive tissue bridging the atrium and ventricle and totally bypassing the *atrioventricular (AV) node*. The Kent bundle is associated with AV reentrant tachycardia and ventricular preexcitation (*Wolff–Parkinson–White-syndrome*).

Late potentials High-*frequency*, low-*amplitude* signals detected at the terminal portion of the signal-averaged QRS, they may represent electrical activity from regions of slow *conduction*.

Left bundle branch The *proximal* portion of the *conduction* system to the left ventricle. Arises from the *His bundle* and divides into an anterior-superior and posterior-inferior branch. When block occurs in this tissue, it results in a wide QRS and is called *left bundle branch block* (LBBB).

Lenègre's disease A degenerative disease of the *cardiac conduction system* due to progressive *fibrosis* that may result in *heart block*.

Long QT syndrome A group of disorders of *repolarization* associated with ventricular arrhythmias, especially *polymorphic ventricular tachycardia* (*torsades de pointes*). Long QT syndrome may be congenital or acquired with causes including antiarrhythmic drugs that affect repolarization.

Lown–Ganong–Levine syndrome A special form of preexcitation syndrome in which there is a short PQ interval with a narrow QRS complex (with no delta wave on the electrocardiogram [ECG]). It has been claimed that the so-called LGL syndrome is a mixture of various supraventricular tachycardias but is not a true syndrome.

Mahaim fiber An accessory *pathway* going directly from the *atrioventricular (AV) node* to either the ventricular *myocardium* or one of the *conduction* fascicles, bypassing the *His bundle*. There may be a small or absent *delta wave* and a very short *PR interval*. This anatomical abnormality predisposes a patient to reentrant arrhythmias.

Mobitz I A type of *atrioventricular (AV) block* in which the *PR interval* progressively lengthens until a *P-wave* appears without an associated *QRS complex*; the PR interval after the pause associated with the non conducted beat is shorter than the PR interval preceding it. The QRS complex may be normal or abnormal. It is sometimes called *Wenckebach* or type I second degree AV block.

Mobitz II A type of *atrioventricular (AV) block* in which the *PR interval* remains constant, but, from time to time, a *P-wave* occurs without an associated *QRS complex*. The QRS complex is usually wide. There are no generally recognized warning signals on the electrocardiogram (ECG) to indicate when a beat will be dropped. Mobitz II may abruptly and without warning revert to ventricular *asystole* and is considered a potentially lethal condition. It is sometimes called type II second degree AV block.

Orthorhythmic pacing A form of *burst antitachycardia* pacing in which the pacing stimuli are coupled to the *intrinsic* tachycardia complexes.

Palpitations Subjective awareness of the heartbeat. While commonly associated with rapid or irregular rhythms, palpitations may also occur in the presence of normal rhythms.

PR interval The period, measured in milliseconds and readily viewed on a surface electrocardiogram (ECG), from the onset of the *P-wave* (atrial depolarization) to the onset of the *QRS complex* (*ventricular depolarization*).

Preexcitation Activation of all or part of the ventricular *myocardium* by an *accessory pathway* at an earlier time than if *conduction* had occurred only through the normal atrioventricular (AV) conduction system. On a surface electrocardiogram (ECG), preexcitation yields a *delta wave* and usually a short *PR interval*. Electrophysiological studies can be used to define the predominant pathway. See also *Wolff–Parkinson–White-syndrome*.

R-on-T A potentially dangerous condition induced when a *premature ventricular contraction* occurs during the *T-wave* of the preceding QRS-T complex. R-on-T phenomenon can induce *ventricular tachycardia* or *ventricular fibrillation*.

Radiofrequency ablation Destruction of tissue with radiofrequency *energy* delivered through an electrode *catheter*.

Rate-responsive ("physiological") pacemakers increase cardiac volume based on biological need. The rate most appropriate to the level of activity is determined by some physiological parameter (biosignals), such as muscle activity, QT interval (catecholamine-dependent intra-cardiac electrogram [EGM] variations), minute ventilation, temperature, pH value, oxygen saturation, pressure, or stroke volume.

Reentry An abnormal *conduction* pathway, either anterograde or retrograde, through which an electrical impulse repeatedly enters and exits. Reentry can cause ectopic contractions, premature atrial depolarizations, supraventricular tachycardias, ventricular tachycardias, and other arrhythmias.

Reentry tachycardia An abnormal condition in which a cardiac impulse may reexcite a region through which it has just traveled. Some *reentry* loops are very large and involve entire *conduction* pathways. Others are microscopic within a small area of tissue. It is also known as circus tachycardia.

Refractoriness A measure of the recovery of *excitability* of cardiac tissue under electrophysiological investigation.

Resting potential The voltage or *potential difference* across a myocardial cell when it is fully repolarized and in the resting state.

Resynchronization therapy (biventricular pacing) A term used to describe the effort to resynchronize the contractions of the lower chambers of the heart (ventricles) by sending tiny electrical impulses to both sides of the heart. Resynchronization can help the heart pump blood more efficiently and has been shown to decrease the symptoms of heart failure in some patients.

Sick sinus syndrome Sick sinus syndrome includes a group of complicated, non-ventricular arrhythmias whose origin is believed to be a disturbance in the function of the heart's natural pacemaker in the sinus node. *Synonyms* for this term are sinus node syndrome and brady-tachy syndrome.

Sinoatrial (SA) block The failure of electrical impulses produced by a normally functioning *sinoatrial node* to reach the atria. There are three degrees of sinoatrial block.

Sinoatrial (SA) node A group of cells that acts as the natural pacemaker in the healthy heart, located high and posterior in the right atrium at the point where the superior vena cava joins the atrial tissue mass. The healthy sinoatrial node initiates 60- 100 beats per minute at rest.

Sinus node recovery time (SNRT) A test of sinus node *automaticity*. The length of time it takes for the sinus node to recover after the discontinuation of sustained atrial pacing. Sinus node recovery time is measured as the longest pause following a period of atrial pacing (usually 30 seconds) at different constant cycle lengths. The sinus node pause following an atrial *output* pulse is caused by overdrive suppression of the sinus node. The normal corrected sinus node recovery time is typically less than 550 ms.

Sinus rhythm Normal cardiac rate set by the sinus node, which in turn is influenced by the *autonomic nervous system*. Normally it is between 60 and 100 beats per minute. It is also called normal sinus rhythm.

Sinus tachycardia A *sinus rhythm* greater than 100 beats per minute.

Stokes–Adams syndrome Paroxysmal heart block or ventricular *dysrhythmia*, producing low or no *cardiac output* resulting in such symptoms as *syncope* or convulsions.

Sudden cardiac death (SCD) A lethal or near-lethal ventricular arrhythmia, most usually *ventricular tachycardia* or *ventricular fibrillation*. The definition varies widely among different investigators, depending upon the maximum amount of time from the onset of symptoms to their conclusion.

Supraventricular tachycardia (SVT) A *tachycardia* that originates outside the ventricle. Supraventricular tachycardias are usually pathological tachycardias, including atrioventricular (AV) and *AV nodal reentry tachycardia* (AVNRT) as well as atrial *fibrillation, atrial flutter,* and ectopic or automatic tachycardias. In common usage, it refers to a reentrant tachycardia. See also *AV nodal reentry tachycardia.*

Sympathetic nervous system Part of the *autonomic nervous system* involved in the control of involuntary bodily functions, including cardiac and blood vessel activity. This system stimulates cardiac activity and produces effects opposite to those of the *parasympathetic nervous system,* which depresses cardiac activity. Some effects of sympathetic stimulation are an increase in *heart rate, cardiac output,* and blood pressure. See also *autonomic nervous system.*

Syncope A fainting or blackout spell caused by insufficient blood supply to the brain.

Tachyarrhythmia (*Automatic* or *triggered impulse formation* and *reentry* [circus movement].) The potentially involved substrate can be not only a preformed, linear conduction pathway, such as the intraventricular conduction system or the accessory pathway between the atrium and ventricle, it could also be the sinus node, atrium, atrioventricular (AV) node, or infarcted or fibrosed ventricular myocardial tissue.

Tachycardia Any cardiac *arrhythmia* characterized by a rapid rate, usually over 100 beats per minute. Tachycardias may be normal, as in the case of a *sinus tachycardia* induced by exercise, or may indicate a pathology.

Temporary (time-restricted) pacemaker therapy with an external pacemaker is indicated for acutely occurring asystole in the case of Adams–Stokes syndrome, cardiogenic (rhythmogenic) shock, or, further, in the case of reversible, paroxysmal conduction disorders with high-degree bradycardia (e.g. myocardial infarction, digitalis toxicity, myocarditis).

Torsade de pointes Polymorphic ventricular tachycardia associated with a long QT interval. Torsades de pointes is so named because it appears that the QRS complexes are twisting around a center axis. The rhythm often self-terminates but tends to recur. Disagreement exists as to whether the underlying mechanism is afterdepolarizations or dispersion of refractoriness.

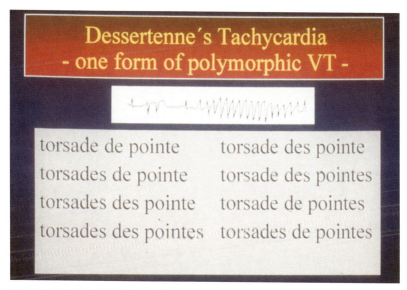

Dessertenne's tachycardia – polymorphically written.

Ventricular fibrillation (VF or VFIB) A chaotic, very rapid ventricular rhythm with disorganized *depolarization* resulting in ineffective contractions, lack of an effective heartbeat, and collapse. If not reversed within a few minutes, irreversible brain damage will result. The most common cause of *sudden cardiac death*, it is amenable to treatment with both external and implantable defibrillators.

Ventricular inhibited (VVI) pacemaker This is a ventricular demand pacemaker system which can pace and sense in the right ventricle (VV) and which delivers an output pulse after a preset or programmed pacing interval expires. When the patient's intrinsic rhythm exceeds the preset pacing rate, or should a premature ventricular contraction occur, the pacemaker senses the QRS complex and inhibits the delivery of an output pulse to the ventricle (I).

Ventricular tachycardia (VT) is a rapid heart rate that starts in the ventricles. During VT, the heart does not have time to fill with enough blood between heartbeats to supply the entire body with sufficient blood. It can be life threatening if it progresses to ventricular fibrillation.

Volt The force with which electrical *current* is driven. *Pulse amplitude* is stated in volts. It is abbreviated as "V."

Wolff–Parkinson–White (WPW) syndrome is characterized by a double stimulation of the ventricles. A premature conduction wave via accessory pathways (*preexcitation*) stimulates the portions of the ventricles nearest the atrium; then the ventricles depolarize as a result of the conduction wave which proceeds normally through the atrioventricular (AV) node. The clinical significance of this syndrome depends on the appearance of (supraventricular) reentry tachycardias.

Part 3

Historical Pages

Hippocrates of Cos (466–ca.370 BC) and the Hippocratic oath*

Hippocrates was indeed an outstanding figure, a renowned personality whose genius has shown in the firmament of Greek civilization and has, up to the present day, guided the scholar and the ordinary man alike. Hippocrates' descent (on his father's side) is traced back to Asclepios, the god of medical science in Greek mythology. Without any technical infrastructure or precision instruments, he succeeded through his intellectual strength alone in discovering fundamental truths and formulating opinions that are, even today, considered to be scientific axioms. With his inquiring mind, his profound knowledge, and his unfailing concern for the sufferer, he consolidated his position as a dazzling beacon in medical science and came to be regarded as the greatest medical

* Lüderitz, B. (2001) History. *Journal of Interventional Cardiac Electrophysiology*, **5**, 119–20 (with permission).

ʼΩφελέειν ἢ μὴ βλάπτειν

(Help, or at least do not harm)

HIPPOCRATES

the oath of hippocrates

Berndt Lüderitz, MD

I swear by Apollo the physician, and Æsculapius, and Hygeia, and Panacea, and all the gods and goddesses, that according to my ability and judgment, I will keep this Oath and its stipulation—to reckon him who taught me this Art equally dear to me as my parents, to share my substance with him, and to relieve his necessities if required; to look upon his offspring in the same footing as my own brothers, and to teach them this Art if they shall wish to learn it, without fee or stipulation; and that by precept, lecture, and every other mode of instruction, I will impart a knowledge of the Art to my own sons, and those of my teachers, and to disciples bound by a stipulation and oath according to the law of medicine, but to none other. I will follow that system of regimen which, according to my ability and judgment, I consider for the benefit of my patients, and abstain from whatever is deleterious and mischievous. I will give no deadly medicine to anyone if asked, nor suggest any such counsel; and in like manner I will not give to a woman a pessary to produce abortion. With purity and with holiness I will pass my life and practice my Art. I will not cut persons laboring under the stone, but will leave this to be done by men who are practitioners of this work. Into whatever houses I enter, I will go into them for the benefit of the sick, and I will abstain from every voluntary act of mischief and corruption; and, further from the seduction of females or males, of freemen and slaves. Whatever, in connection with my professional practice, or not in connection with it, I see or hear, in the life of men, which ought not to be spoken of abroad, I will not divulge, as reckoning that all such should be kept secret. While I continue to keep this Oath unviolated, may it be granted to me to enjoy life and the practice of this Art, respected by all men, in all times. But should I trespass and violate this Oath, may the reverse be my lot.

genius in history. He was the first to apply the eternal laws of nature to scientific research, and to teach that experience, observation, and experiment are the most reliable guides for the doctor and, more generally, the natural scientist. In this way he raised medicine from the status of a multiform empirical art to that of a uniform, systematic science: "God is not the cause of anything (evil)." (Hippocrates)

In the treatise *On Auscultation and the Invention of the Stethoscope*, the famous scholar René Théophile Hyacinthe Laënnec (1781–1826) categorically confirms that his method was inspired by Hippocrates.

Hippocrates is regarded as a rationalist; however, his imperative command, "Help or at least do not harm" is today the highest ideal of all great scientists and the foundation of humanistic medicine. Hippocrates will always teach us that, ". . . where there is love of men, there is also love of the art." (Hippocrates)

The first of the *aphorisms*, which is considered to be the most famous of the genuine Hippocratic texts, has become proverbial: "Life is short and the art is long; the opportunity fleeting, experiment dangerous, and judgement difficult." (Hippocrates)

The oath of Hippocrates influenced much more than any other scientific ethics as it radiates absolute respect for man and disinterested love to science. The Hippocratic oath concerns not only doctors but all scientists and professional people who believe in humanism and regard their profession as a sacred mission in society. Within the space of just one page, one can find eternal truths, enduring principles, imperishable precepts, and admonitions that are eternally valid. Hippocrates' year of death cannot be precisely calculated, but it can be placed between the years 370 and 358 BC.

References
Kiapokas, M.S. (1999) *Hippocrates of Cos and the Hippocratic Oath*. Cultural Center, Cos.
Lüderitz, B. (1998) *History of the Disorders of Cardiac Rhythm*, 2nd edn. Futura, Armonk, NY.

Marcus Gerbezius (1658–1718)*

Marcus Gerbezius was born in Slovenia (Šentvid near Stična not far from Ljubljana) on October 24, 1658. Upon completing his study of philosophy at the University of Laibach (today Ljubljana), he studied medicine at the universities in Vienna, Padua, and Bologna, and

---

* Lüderitz, B. (2002) History. *Journal of Interventional Cardiac Electrophysiology*, **6**, 96 (with permission).

Logo of the Slovenian Society of Cardiology.

graduated from Bologna in 1684. Gerbezius then commenced his medical practice in Carniola and Ljubljana. In 1689, he became a member of the renowned German Academy of Natural Scientists in Halle (Academia Cacsarea Leopoldina—Carolina Naturae Curiosorum). From the start of his membership in the Academy of Natural Scientists until his death in Ljubljana on March 9, 1718, Marcus Gerbezius published a great number of his medical observations in the Academy's periodicals.

In 1717, based on an extremely accurate pulse analysis, Gerbezius described the symptoms of bradycardia most probably induced by complete atrioventricular (AV) block. However, these observations were not published until 1718 (posthumously) in the book *Constitutio Anni 1717* AD. *Marco Gerbezio Labaco 10. Decem. descripta. Miscellanea-Ephemerides Academiae Naturae Curiosorum. Cent. VII, VIII. (1718); in Appendice.*

Gerbezius's descriptions preceded those of Giovanni Morgagni by 44 years. In fact, Morgagni mentions Gerbezius several times in his work *De Sedibus et Causis Morborum per Anatomen Indagatis* when referring to the characteristics of the pulse, symptoms, and the course of the disease in a patient with AV block. Hence, it could be suggested that the Morgagni–Adams–Stokes syndrome could be known as the *Gerbezius–Morgagni–Adams–Stokes* syndrome!

References

Cibic, B. & Kenda, M.F. (2000) First description of syncopal attacks in a patient with a heart block. Description of Marko Gerbecius' contribution to the symptomatology of "Morgagni–Adams–Stokes syndrome." Eighth Alpe Adria Cardiology Meeting, Portorož, Slovenia.

Mušič, D. (1977) *Marko Gerbec/Marcus Gerbezius 1658–1718. Syndroma Gerbezius–Morgagni–Adams–Stokes.* Ljubljana.

Mušič, D., Rakovec, P., Jagodic, A. & Cibic, B. (1984) The first description of syncopal attacks in heart block. *Pacing and Clinical Electrophysiology: PACE,* **7**, 301–3.

Paracelsus (1493–1541)—Stormy petrel of medicine*

In the Renaissance "chemical kitchens" of Aureolus Philippus Theophrastus Bombastus von Hohenheim, who boastfully called himself Paracelsus, many things were brewed. chemicals, polypharmacol mixtures, serious medical writings; and vitriolic, abusive attacks upon medical colleagues, religionists, and political officials. Swiss-born Paracelsus' controversies forced him to travel widely and to move frequently. Labeled genius by some, quack by others, his medical efforts got results, and patients liked him. He attacked medieval "sacred cows," Galen and

---

* Lüderitz, B. (1997) History. *Journal of Interventional Cardiac Electrophysiology,* **1**, 163 (with permission).

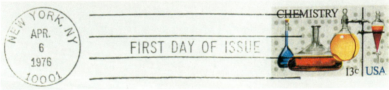

Avicenna, and helped turn medicine from them to rational research. He attempted to manufacture new remedies, and he advocated the use of chemicals in medicine.

Everything is a poison
the dose alone makes
a thing not a poison.
(Paracelsus, August 19, 1538)

Jan Evangelista Purkinje (1787–1869)*

Jan Evangelista Purkinje was born on December 17, 1787 in Libochovice, Bohemia, and graduated in medicine from Prague. While he was

---

* Lüderitz, B. (1998) History. *Journal of Interventional Cardiac Electrophysiology*, **2**, 391 (with permission).

Professor of Physiology at the University of Wroclaw (Breslau), he made several important scientific discoveries. He was one of the very first to use the microscope to explore the function of tissues, and he introduced the term protoplasm. In 1839, he described the subendocardial structures in the heart, known ever since as the "Purkinje fibers." In 1850, Purkinje went to Prague as a Professor of Physiology, where he likewise founded a physiology institute, of which he was the director until his death on July 28, 1869.

Reference
Purkinje, J.E. (1845) Mikroskopisch-neurologische Beobachtungen. *Arch Anat Physiol Wiss Med*, **II/III**, 281–95.

Carl Friedrich Wilhelm Ludwig (1816–1895)*

Carl Friedrich Wilhelm Ludwig was born in Witzenhausen, Germany, on December 29, 1816. He studied at several universities including Marburg, Erlangen, and Hamburg. Following graduation, he was

Carl Friedrich Wilhelm Ludwig (1816–1895). German physiologist who made numerous contributions to cardiovascular physiology and invented the kymograph.

* Lüderitz, B. (2004) History. *Journal of Interventional Cardiac Electrophysiology* 11, 219–200 (with permission).

The effigy of Carl Ludwig on the honorary medal named after him—the outstanding pioneer and representative of modern science in the nineteenth century. This medal is the highest award of the German Cardiac Society (GCS). His emblem can be found on the program of the annual meetings of the GCS demonstrating the significance of physiological and pathophysiological topics in the framework of the society.

appointed Assistant in Anatomy under Ludwig Fick at the University of Marburg. In 1865, he was appointed Professor of Physiology at the University of Leipzig. Ludwig designed a new physiologic institute at Leipzig which was the most advanced experimental laboratory in the world when it opened in 1869. Ludwig encouraged many of his advanced pupils to investigate the physiology of the heart and circulation, among them Henry P. Bowditch from the USA; Robert Tigerstedt, Luigi Luciani, Karl Ewald Hering, Adolf Fick, and Otto Frank from the European continent; and also Augustus D. Waller from the UK.

Ludwig's interest in the circulatory system lead him to develop an instrument to record hemodynamic and other physiologic events accurately. For example, by simultaneously recording the pulse wave and respiratory pattern, he first described sinus arrhythmia in 1847. The first registration of ventricular fibrillation is depicted on p. 144.

To quote the pioneering Scottish pharmacologist T. Lauder Brunton, "More than to anyone else since the time of Harvey, do we owe our present knowledge of the circulation to Carl Ludwig." (T. Lauder Brunton)

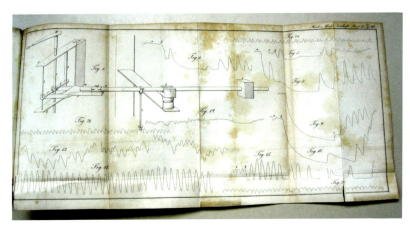

First graphic documentation of ventricular fibrillation: In 1849, while investigating vagal influences on cardiac activity, M. Hoffa, in Carl Ludwig's laboratory, documented bizarre unregulated actions of the ventricles when exposed directly to strong faradic or constant currents. The disorder affected both rhythm and intensity, persisted after termination of electroexcitation and stopped cardiac output. The atria did not participate in the arrhythmia.

References

Fye, W.B. (2003) Carl Ludwig. In: *Profiles in Cardiology*. (eds J.W. Hurst, C.R. Conti & W.B. Fye), pp. 128–9. Foundation for Advances in Medicine and Science, Mahwah, NJ.

Hoffa, M. & Ludwig, C. (1850) Einige neue Versuche über Herzbewegung. *Zeitschrift Rationelle Medizin*, **9**, 107–44.

Lüderitz, B. (2002) *History of the Disorders of Cardiac Rhythm*, 3rd edn. Futura, Armonk, NY.

Lüderitz, B. & Arnold, G. (eds) (2002) German Cardiac Society. *German Journal of Cardiology*, **91** (Suppl. 4), 26.

Etienne Jules Marey (1830–1904)*

Etienne Jules Marey was born March 5, 1830 in Beaune, Burgundy (Côte d'Or), France. His father was an assistant to a Burgundy wine merchant. Marey's thesis, completed at the age of 29 years, dealt with the circulation of the blood under normal and pathological conditions. This work marks the beginning of his studies on cardiovascular hemodynamics.

---

* Lüderitz, B. (2004) History. *Journal of Interventional Cardiac Electrophysiology* (in press) (with permission).

E.J. Marey

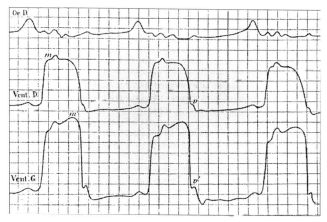

Pressure tracings as obtained by E.J. Marey. Variations in pressure within the atria and ventricles during the cardiac cycle of a horse. *Top*: right atrium; *middle*: right ventricle; *bottom*: left ventricle.

His main interest, however, was engineering, particularly focused on graphic registration of changes in time and place; especially concerning movement of the heart and pulsation of the arteries or locomotion of animals—in the air, in the water, and on the ground—and humans. Marey's interest was centered around movement in all its forms: cardiovascular hemodynamics, respiration, muscular contraction, and complex movements. For this he invented many scientific instruments himself. Together with Auguste Chaveau, a Professor of Veterinary Physiology in Lyon, he described and interpreted: simultaneous pressures in the right atrium and right ventricle, and the left ventricle and aorta; the pulmonary artery pressure; the atrial influence on the ventricular pressure curve; the isometric phase of ventricular contraction; the chronology of valve motion; and the synchrony of left and right ventricular contraction. Other original work on the heart by Marey included the discovery of the refractory period of heart muscle in 1875, and the first recording in animals of the electrogram of the heart using a capillary electrometer in 1876, preceding the work of A.D. Waller who recorded the first human electrogram in 1887. The graphic registration was replaced after 1881 by photographic records. Marey contributed considerably to the development of this new tool of registration and evaluation of movement ("chronophotography"), which finally resulted in the invention of the cinematograph. Furthermore, Marey's studies on the flight of birds lead him to research on gliding and aviation. Marey, who was a professor at the Collège de France and a member of the Academy of Science, produced numerous scientific papers, drawings, photographs, and films (the first in the history of cinema!). Etienne Jules Marey, a technical genius, the "engineer of life" and inventor of cinematography died on May 15, 1904 at his Paris home.

References
Régnier, C. (2003) Etienne Jules Marey, the "engineer of life." *Medicographia*, **25**, 268–74.
Silverman, M.E. (2003) Etienne Jules Marey: Nineteenth century cardiovascular physiologist and inventor of cinematograph. In: *Profiles in Cardiology*. (eds J.W. Hurst, C.R. Conti & W.B. Fye). Foundation for Advances in Medicine and Science, Mahwah, NJ. pp. 143–5.
Snellen, H.A. (1980) Introduction to the exhibition E. J. Marey 1830–1904 on the occasion of the Eighth European Congress of Cardiology. Rotterdam: Kooyker Scientific Publications.

Augustus Desiré Waller (1856–1922)—The first to record the electrical activity of the human heart*

Augustus Desiré Waller was born in Paris on July 12, 1856, the son of the celebrated physiologist Augustus Volney Waller. After studying

---

* Lüderitz, B. (2003) History. *Journal of Interventional Cardiac Electrophysiology*, **9**, 59–60 (with permission).

A.D. Waller in his laboratory with laboratory dog "Jimmy."

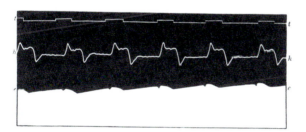

The first human electrocardiogram (ECG), recorded by A.D. Waller in 1897.

medicine at the universities in Aberdeen and Edinburgh in Scotland and Leipzig in Germany, Waller was made a professor in Aberdeen in 1881. He worked first in the physiology laboratory in London, UK, under Professor John Burdon–Sanderson and gave lectures on physiology at the London School of Medicine for Women. Waller held the same position for 16 years at the Medical School of St. Mary's Hospital. He was named Director of the Physiology Laboratory of the University of London in 1902. Waller died in London on March 11, 1922.

Waller primarily studied the electrical phenomena of the heart. As early as 1887, he was able to obtain an electrocardiogram (ECG) from

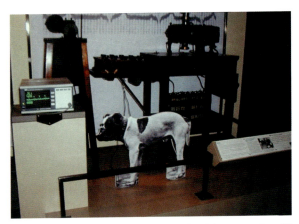

Waller's table for electrogram recording behind his laboratory dog "Jimmy" (and a modern electrocardiogram [ECG] monitor). Courtesy of The Bakken. A Library and Museum of Electricity in Life. Minneapolis, MN, USA.

the body surface of a human being with the aid of a Lippmann capillary electrometer. Although the clinical significance of this ECG was not recognized at the time, Waller's work nevertheless laid the foundation for modern electrocardiography.

Waller found that electrical currents generated by the heart could be recorded with a mercury capillary eletrometer when the electrodes were placed on the chest or the limbs. The capillary electrometer was devised in 1873 by Gabriel Lippmann (1845–1921): it consists of a glass tube containing mercury with one end drawn out into a fine capillary (20–30 μm) and immersed vertically in dilute sulfuric acid. Measurement is based on displacement of the mercury meniscus because mercury contracts and expands according to the potential difference between the mercury and acid which are connected to electrodes on two points on the body.

Waller's classic demonstration of the human ECG (called the electrogram at the time already) from the intact human heart took place at St. Mary's Hospital, London, in May 1887, with surface electrodes strapped to the front and back of the chest. There were only two distorted deflections: ventricular depolarization and repolarization. The P-wave was not discernible with the 1887 apparatus. This historic event in 1887 was also witnessed by W. Einthoven. The following year, Waller recorded the ECG by using saline jars in which the extremities were immersed. Einthoven, himself, credited Waller with the first human ECG.

References
Barold, S.S. (2002) A short history of electrocardiography 100 years after Einthoven's first recording. In: *Einthoven 2002. 100 Years of Electrocardiography.* (eds M.J. Schalij, M.J. Janse, A. van Oosterom, E.E. van der Wall & H.J. Wellens), p. 171. The Einthoven Foundation, Leiden, the Netherlands, pp. 171–8.
Lüderitz, B. (2002) *History of the Disorders of Cardiac Rhythm*, 3rd edn. Futura, Armonk, NY.

Willem Einthoven (1860–1927)—A hundred years of electrocardiography*

Willem Einthoven was born on May 21, 1860, a son of a military doctor in Semarang on the island of Java. After the death of his father, the family returned to the Netherlands in 1870, where Einthoven completed high school and started medical school at the University of Utrecht in 1879. There he earned his doctorate in 1885. That same year, he was granted a position as Professor of Physiology and Histology at the University of Leiden, even though he had not yet completed his final examinations. Einthoven held that position until his death on September 28, 1927.

After Waller had succeeded in 1887 in recording the first electrocardiogram (ECG) from the body surface of a human, Einthoven began his work in 1895 with the Lippmann capillary electrometer. In consideration of physical factors, Einthoven corrected the coarsely-differentiated

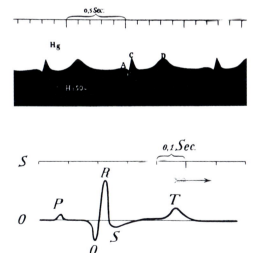

* Lüderitz, B. (1999) History. *Journal of Interventional Cardiac Electrophysiology*, 3, 353 (with permission).

Dutch stamp commemorating the ninetieth anniversary of the first registration
of an modern electrocardiogram (ECG) by Willem Einthoven (1860–1927).

capillary image, which led him to refine the string galvanometer, invented
in 1897 by C. Ader. In a proclamation issued in 1901 to commemorate the
70th birthday of the Dutch scholar Bosscha, Einthoven described his
work in refining the galvanometer and gave broad access to his work by
publishing *A New Galvanometer* in 1903 for reference, see below.

References

Ader, C. (1897) Sur un nouvel appareil enregistreur pour cables sous-marins.
   Comptes Rendus Hobdomadaires des Seances de L'Academie des Sciences, **124**, 1440–2.
Einthoven, W. (1903) Die galvanometrische Registrierung des menschlichen
   Elektrokardiogramms, zugleich eine Beurteilung der Anwendung des Kapillar-
   Elektrometers in der Physiologie. *Pflügers Archiv*, **99**, 472–80.
Waller, A.G. (1887) A demonstration on man of electromotive changes accom-
   panying the heart's beat. *The Journal of Physiology*, **8**, 229–34.

I.C. Brill; Tachycardia-related cardiomyopathy*

The hemodynamic consequences of atrial fibrillation (AF) depend on
numerous issues. Among these, the heart rate is of major concern. It

---

* Lüderitz, B. (1999) History. *Journal of Interventional Cardiac Electrophysiology*, **3**,
287 (with permission).

Am Heart J 1937; 13:175-182

AURICULAR FIBRILLATION WITH CONGESTIVE FAILURE AND NO OTHER EVIDENCE OF ORGANIC HEART DISEASE*.

I. C. BRILL, M.D.

PORTLAND, OREGON

IN RECENT years it has become generally recognized that auricular fibrillation may occur in an otherwise normal heart. Among the more important publications relating to this problem are those of Parkinson and Campbell,[1] Fowler and Baldridge,[2] Friedlander and Levine,[5]

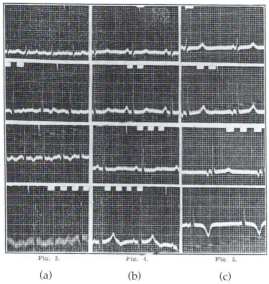

Fig. 3. Fig. 4. Fig. 5.

(a) (b) (c)

(a) Electrocardiogram (ECG) taken June 17, 1935, showing auricular fibrillation and some evidence of digitalis effect. Ventricular rate, 110. (b) ECG taken June 19, 1935, showing sinus rhythm and some evidence of digitalis effect. Ventricular rate 75. (c) ECG taken July 11, 1935, showing normal tracing except for inverted P-waves in leads II and III. Ventricular rate, 75. Courtesy of (1937) *American Heart Journal*, **13**, 175–82.

is well recognized that chronic tachycardia in patients with AF and fast ventricular response may result in extensive changes in ventricular function and structure referred to as tachycardia-induced cardiomyopathy (Gallagher, 1985; Fenelon *et al.*, 1996; Schumacher & Lüderitz, 1998). These consequences of tachyarrhythmia are crucial since it may be reversible with tachycardia rate or rhythm control. The first clinical description of chronic uncontrolled tachycardia resulting in a reversible left ventricular dysfunction was as early as 1937 by I.C. Brill. This investigator concluded from his observations in patients with AF that, "auricular fibrillation, apart from any other disease of the heart, may cause severe congestive failure and that upon cessation of the arrhythmia the congestive failure may be followed by complete and lasting recovery" (Brill, 1937). By now, multiple studies in animal models and clinical investigations have supported this hypothesis.

References

Brill, I.C. (1937) Auricular fibrillation with congestive failure and no other evidence of organic heart disease. *American Heart Journal*, **13**, 175–82.

Fenelon, G., Wijn, W., Andries, E. & Brugada, P. (1996) Tachycardiomyopathy: mechanisms and clinical implications. *Pacing and Clinical Electrophysiology: PACE*, **19**, 95–106.

Gallagher, J.J. (1985) Tachycardia and cardiomyopathy: The chicken–egg dilemma revisited. *Journal of the American College of Cardiology*, **6**, 1172–3.

Schumacher, B. & Lüderitz, B. (1998) Rate issues in atrial fibrillation: Consequences of tachycardia and therapy for rate control. *The American Journal of Cardiology*, **82**, 29N–36N.

Karel Frederik Wenckebach (1868–1940)*

Karel Frederik Wenckebach, a Dutch-born physician, famous for his first description of premature heartbeats in the last decade of the nineteenth century, studied physiology under T.W. Engelmann at the University of Utrecht, the Netherlands. In 1914, he became Director of the I. Department of Medicine at the University of Vienna, Austria, where he lived until his retirement in 1929; he remained in Vienna until his death in 1940.

Under his leadership, the I. Department of Medicine became famous worldwide in those years. Wenckebach revolutionized our under-

* Lüderitz, B. (2003) History. *Journal of Interventional Cardiac Electrophysiology*, **8**, 77 (with permission).

K.F. Wenckebach

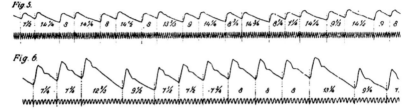

Fig. 5.

Fig. 6.

Sphygmographic registration of the radial pulse. Second-degree atrioventricular (AV) block—Wenckebach block (Mobitz I) Each successive (atrial) impulse encounters a longer and longer delay in the AV node until one impulse fails to make it through.

standing of cardiology, not only with his first descriptions of premature beats but also with his standard work entitled *Die unregelmässige Herztätigkeit und ihre klinische Bedeutung* which was published for the first time in 1914, and revised and republished in 1927 (Wenckebach, 1899; Wenckebach, 1914).

He also initiated the use of quinidine as an antiarrhythmic drug, which he administered for the first time to a patient with atrial fibrillation in 1914. After 1 g of quinine, sinus rhythm was restored (Wenckebach, 1923).

He was famous for his lectures about percussion and auscultation, but nevertheless (since Wenckebach was a boring lecturer) a rhyme became very popular at that time among the medical students: *"Beim Herrn Professor Wenckebach, sind nur die ersten Bänke wach!"* ("When listening to Wenckebach, only the first row can stay awake!")

References

Lüderitz, B. (2002) *History of the Disorders of Cardiac Rhythm*, 3rd edn. Futura, Armonk, NY.

Lüderitz, B. & Barold, S.S. (2002) Woldemar Mobitz. (1889–1951). History. *Journal of Interventional Cardiac Electrophysiology*, **7**, 261.

Wenckebach, K.F. (1899) De Analyse van den onregelmatigen Pols. II. Over eenige Vormen van Allorhythmie en Bradycardie. *Nederl Tijdschr Geneesk*, **35**, 665.

Wenckebach, K.F. (1914) *Die unregelmässige Herztätigkeit und ihre klinische Bedeutung*. Engelmann, Leipzig, Germany.

Wenckebach, K.F. (1923) Cinchone derivatives in the treatment of heart disorder. *JAMA: The Journal of the American Medical Association*, **81**, 472.

Woldemar Mobitz (1889–1951)*

Woldemar Mobitz's name became known through his fundamental article in 1924 on partial atrioventricular (AV) block and its classification. Mobitz was born on May 31, 1889 in St. Petersburg, Russia. He attended the local high school in Meiningen (Saxony, Germany) from

The University Hospital Freiburg with the monument of Adoph Kussmaul (1909) where W. Mobitz was Associate Professor. The hospital was destroyed 1944.

Diagram of type II, second degree atrioventricular (AV) block by Mobitz in 1924. Note the occurrence of a 3 : 2 AV block in the middle of the diagram and the constant AV conduction before and after the block. The sinus rate is constant and the ventricular intervals are exact multiples of the sinus interval because AV conduction and the sinus rate remain unchanged.

---

* Lüderitz, B. (2002) History. *Journal of Interventional Cardiac Electrophysiology*, 7, 261 (with permission).

which he graduated in 1908. He then studied medicine at the Universities of Freiburg and Munich, where he earned his doctorate in 1914 ("Contributions to Basedow disease"). He then worked at the Surgical Hospitals in Berlin and Halle as well as in internal medicine at the University Hospitals of Munich and Freiburg. In Munich, Mobitz was promoted to the position of a senior lecturer thanks to his research on heart block. In 1928, after a 4-year tenure, he accepted a post in Freiburg as Associate Professor and Chief of Staff of the Clinic of Internal Medicine. In 1943, he became Director of the Medical Hospital in Magdeburg-Sudenburg Municipal Hospital until the occupation by the Soviet army in 1945.

Mobitz's work was devoted to internal medicine and he was especially interested in cardiology. From 1924 to 1928, he published his famous key papers on AV dissociation and heart block.

In 1924, Mobitz differentiated two types of second degree AV block with the aid of the electrocardiogram (ECG) and characterized their prognostic significance. With type I (Mobitz type I), the PQ interval increases gradually until there is a breakdown of AV conduction. This form is identical to the previously described type of second-degree AV block by Wenckebach at the end of the nineteenth century. With type II block (Mobitz type II), all conducted beats show a constant, typically normal PQ interval, and conduction to the ventricles occurs at regular intervals. This form is identical to the type of AV block described by Hay in 1906 without the benefit of electrocardiography. Mobitz included 2 : 1, 3 : 1 AV block in his type II classification, and indicated the serious nature of type II block and its propensity to Stokes–Adams attacks.

References

Hay, J. (1906) Bradycardia and cardiac arrhythmia produced by depression of certain of the functions of the heart. *Lancet*, **1**, 139–43.

Kastor, J.A. (2000) Arrhythmias, 2nd edn, p. 35. W.B. Saunders, Philadelphia, PA.

Lüderitz, B. (2002) *History of the Disorders of Cardiac Rhythm*, 3rd edn. Futura, Armonk, NY.

Mobitz, W. (1924) Über die unvollständige Störung der Erregungsüberleitung zwischen Vorhof und Kammer des menschlichen Herzens. *Zeitschrift für die gesamte experimentelle Medizin*, **41**, 180–237.

Mobitz, W. (1928) Über den partiellen Herzblock. *Zeitschrift für klinische Medizin*, **107**, 449–62.

Sunao Tawara (1873–1952) A macroscopic image of the left ventricle of the human heart*

Sunao Tawara, the famous Japanese pathologist was born on July 5, 1873 in Oita, Japan; he died 1952. The associated eponym is: Tawara's node: The atrioventricular node which is the beginning of the auricular-ventricular bundle of His.

Sunao Tawara studied at the Imperial University in Tokyo, graduating there in 1901, Igaku Hakushi 1908. The year 1903 to 1906 he spent in Marburg studying pathology and pathological anatomy with Karl Albert Ludwig Aschoff (1866–1942). It was here he undertook his important works on the anatomy and pathology of the heart. When returning to Japan he was appointed extraordinary professor of pathology in Fukuoka, becoming ordinarius of this specialty in 1908.

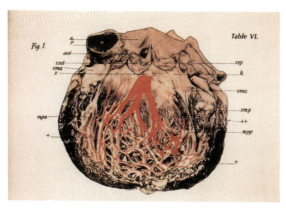

The specific conduction system in the left ventricle of the human heart. The left ventricle is on the anterior wall between the two papillary muscles, opening from the aortic ostium to the apex and opened on both sides. a, aorta; acd, right coronary artery; k, node, that is, the atrial portion of the atrioventricular (AV) node; mpa, anterior papillary muscle; mpp, posterior papillary muscle; P, pulmonary artery; vma, anterior mitral valve; vmp, posterior mitral valve; vsd, right aortic valve; vsp, posterior aortic valve; X, point of division between the connective bundle of the right and left branches; +, terminal extensions of the connective bundle; + +, a strand about 2 cm long and as thick as a horse hair, running from the tip of the posterior papillary muscle upwards through the ventricle and joining the upper posterior section of the ventricular septum. This strand leads a small section of the left branch backwards from the papillary muscle toward the previously mentioned portion of the ventricular wall. From Sunao Tawara (1906) *Das Reizleitungssystem des Säugetierherzens*. Gustav Fischer Verlag, Jena, Germany.

* Lüderitz, B. (1997) History. *Journal of Interventional Cardiac Electrophysiology*, **1**, 79 (with permission).

Paul Dudley White (1886–1973)*

Paul Dudley White was born in Roxbury, Massachusetts, on June 6, 1886. After attending Roxbury Latin School, White attended Harvard College from 1904, entering Harvard Medical School in 1907. He obtained his BA degree in 1908 and graduated from medical school in 1911. He then entered residency at Massachusetts General Hospital. A 1-year study trip to England in 1913 to work with Thomas Lewis (1881–1945) familiarized White with the techniques of electrocardiography. Upon returning to Massachusetts General Hospital, White built the first electrocardiogram (ECG) laboratory outside of England in 1914 and worked on the practical application of ECG recording. World War I interrupted his research; he served as a military physician and then was sent to Greece as a member of the American Red Cross. When he returned to Massachusetts, he served primarily in the capacity of Chair for the cardiology training program for medical students, scientists, and clinicians. He cofounded the American Heart Association, and published in 1931 a book, which had been prepared in 1928, called *Heart Diseases*, (New York; The Macmillan Co, 1951). This book, considered a classic in

First day of issue of the commemorative stamp honoring the one-hundredth birthday of P.D. White, released in honor of the Tenth World Congress of Cardiology, held in Washington, D.C. in 1986.

---

* Lüderitz, B. (1999) History. *Journal of Interventional Cardiac Electrophysiology*, **3**, 193 (with permission).

cardiology, earned White his reputation as a cardiologist. His reputation was further enhanced by numerous other publications, including a classification of cardiology diagnosis, "soldier's heart syndrome" or "effort syndrome," and the first published reports of Wolff–Parkinson–White syndrome in 1930. He gained international repute when numerous scientific trips took him to such places as the former USSR and China. White was also the cofounder of numerous medical organizations. Among other things, he created the International Society of Cardiology, of which he was President from 1954 to 1958. Paul Dudley White died in Boston on October 31, 1973.

References
Dimond, E.D. (1965) Paul Dudley White. A portrait. *The American Journal of Cardiology*, **15**, 434–552.
Lüderitz, B. (1998) *History of the Disorders of Cardiac Rhythm*, 2nd edn. Futura, Armonk, NY.
Wolff, L., Parkinson, J. & White, P.D. (1930) Bundle-branch block with short PR interval in healthy young people prone to paroxysmal tachycardia. *American Heart Journal*, **5**, 685–704.

Wolff–Parkinson–White syndrome: Louis Wolff (1898–1972), John Parkinson (1885–1976), Paul Dudley White (1886–1973)*

Wolff–Parkinson–White (WPW) syndrome was first described in 1930 by L. Wolff, J. Parkinson, and P.D. White as bundle branch block with short PR interval in healthy young people prone to paroxysmal tachycardia.

---

* Lüderitz, B. (1997) History. *Journal of Interventional Cardiac Electrophysiology*, **1**, 255 (with permission).

The American Heart Journal

Vol. V   August, 1930   No. 6

Original Communications

BUNDLE-BRANCH BLOCK WITH SHORT P-R INTERVAL IN HEALTHY YOUNG PEOPLE PRONE TO PAROXYSMAL TACHYCARDIA

Louis Wolff, M.D., Boston, Mass., John Parkinson, M.D., London, Eng., and Paul D. White, M.D., Boston, Mass.

ABERRANT ventricular complexes of the type generally recognized as indicating bundle-branch block were first produced by Eppinger and Rothberger,[3, 4] by the experimental division of the right branch of the His bundle. Eppinger and Stoerk[5] observed similar curves in five patients, and at autopsy demonstrated division of the right branch of the His bundle in two of these. The work of Cohn and Lewis,[2] and of Carter[1] indicated, however, that gross lesions of the main branches are not invariably found at autopsy in patients who during life present this type of electrocardiogram.

Following these original contributions to the subject, bundle-branch block curves have been observed as a temporary sign during infections, congestive failure, coronary thrombosis, tachycardias, and various toxic states. In most if not in all of the reported cases the abnormal curves occurred in patients with definite structural heart disease, or with extreme tachycardia. The references already cited[1, 2] indicate that the type of electrocardiogram under discussion may be obtained in the absence of gross division of a bundle branch.

Experimentally, bundle-branch block curves may be obtained in normal hearts by causing an impulse to enter one bundle branch later than the other. The same result would be produced should the impulse travel through the bundle branches at different speeds or by an aberrant course. That such a mechanism may occur in human beings with normal hearts seems likely from a study of the cases described in the present paper, the presumption being that vagal stimulation may, in certain individuals, give rise to aberrant ventricular complexes.

We have observed the occurrence of bundle-branch block curves in healthy young adults and children with apparently normal hearts. The curves may be typically those of complete right or left bundle-branch block, or of intermediate or lesser grades of aberration. Spon-

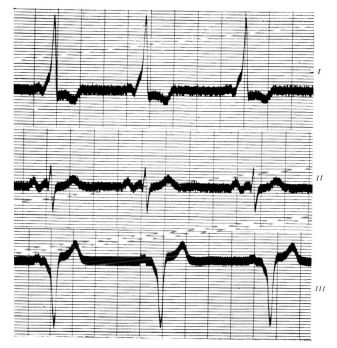

Right bundle branch block. The PR interval is well under 0.1 second. The rate varies between 60 and 70 bpm. From Wolff, L., Parkinson, J. & White, P.D. Bundle branch block with short PR interval in healthy young people prone to paroxysmal tachycardia. *American Heart Journal* 1930, **5**, 685, with permission.

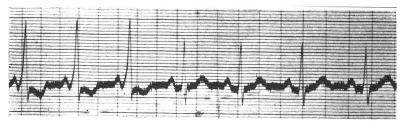

Spontaneous reversion from bundle branch block curves to normal ones. The form of the P-wave remains unaltered, but the PR interval changes from 0.09 second to 0.15 second. From Wolff, L., Parkinson, J. & White, P.D. Bundle branch block with short PR interval in healthy young people prone to paroxysmal tachycardia. *American Heart Journal* 1930, **5**, 685, with permission.

Franz Maximilian Groedel (1881–1951)*

Franz Maximilian Groedel was born on May 23, 1881 in Bad Nauheim, Germany. He was a pioneer of electrocardiography, cardiac radiology, and scientific hydrotherapy, and—most importantly—he was the founder

* Lüderitz, B. (2002) History. *Journal of Interventional Cardiac Electrophysiology*, **6**, 197 (with permission).

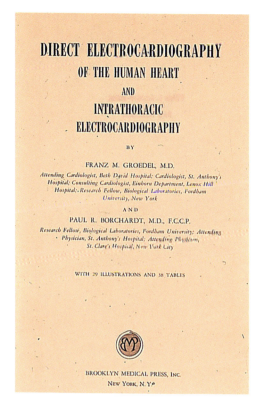

DIRECT ELECTROCARDIOGRAPHY
OF THE HUMAN HEART
AND
INTRATHORACIC
ELECTROCARDIOGRAPHY

BY

FRANZ M. GROEDEL, M.D.
Attending Cardiologist, Beth David Hospital; Cardiologist, St. Anthony's Hospital; Consulting Cardiologist, Einhorn Department, Lenox Hill Hospital; Research Fellow, Biological Laboratories, Fordham University, New York

AND

PAUL R. BORCHARDT, M.D., F.C.C.P.
Research Fellow, Biological Laboratories, Fordham University; Attending Physician, St. Anthony's Hospital; Attending Physician, St. Clare's Hospital, New York City

WITH 29 ILLUSTRATIONS AND 38 TABLES

BROOKLYN MEDICAL PRESS, INC.
NEW YORK, N.Y.

of the American College of Cardiology. In 1904, he received his medical degree from the University of Leipzig in Germany. Groedel cofounded the German Society of Heart and Circulation Research in 1924. His main interests included clinical electrocardiography. He developed the concept of the unipolar chest lead or precordial electrode in the early 1930s independent of Frank Wilson's group at the University of Michigan. Groedel summarized two decades of electrocardiographic research in a 1934 book that included his controversial theory that each cardiac ventricle generated an independent or "partial" electrocardiogram (ECG). Groedel's later work concerned studies on the direct recording of the ECG from the surface of the heart during surgery in humans, particularly from the surface of the atria and ventricles. By 1932 Groedel had published nearly 300 scientific articles, was a full professor at the University of Frankfurt, had a successful practice in Bad Nauheim, and was director of a world class cardiovascular research institute there. In 1933, as Hitler came into power, Groedel was classified as "non-Aryan," and he knew that his career and his life were at risk. Groedel immigrated to the USA in the same year. He died October 12, 1951 in New York.

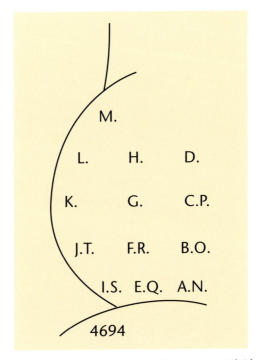

Diagram demonstrating the contact points for the tracings with identical figures from Table 2. A to M, contact places at the anterior wall, N to T at the posterior wall.

Dirk Durrer (1918–1984)*

Dirk Durrer was appointed Professor of Cardiology and Clinical Physiology at the University of Amsterdam, the Netherlands, from 1957 until his death in 1984. In the early 1960s he introduced the famous multiterminal intramural needle electrode. This might be considered as the beginning of clinical electrophysiology, setting the stage for programmed electrical stimulation and registration in the human heart. Durrer's experimental and clinical research gave the electrophysiologic community important new insights in the reentry (circus movement) concept as an explanation of certain types of tachycardias, particularly in the Wolff–Parkinson–White syndrome. Furthermore, Durrer initiated clinical and scientific approaches to pharmacological treatment of life-threatening tachyarrhythmias.

---

* Lüderitz, B. (2000) History. *Journal of Interventional Cardiac Electrophysiology*, **4**, 547 (with permission).

The text on the flagstone *"primum movens, ultimum moriens"* is the title of the inaugural lecture by Durrer in 1957. It means, "What moves first, dies last," and has reference to the functioning of the heart.

In honor of his outstanding work, a monument was erected for Dirk Durrer in 1986 at the Minerva Plein in Amsterdam near to his house in the Rubensstraat 27; Her Royal Highness Princess Juliana, the former Queen of the Netherlands, unveiled on May 28th 1986, this monument which (probably) shows a model with two parallel atrioventricular connections: the specialized conduction pathway and the accessory pathway; and so shows and supports the "reentry theory."

References

Durrer, D., van Dam, R.Th. & Freud, G.E. *et al.* (1970) Total excitation of the isolated human heart. *Circulation*, **XLI**, 899–912.

Durrer, D., Schoo, L., Schuilenburg, R.M. & Wellens, H.J.J. (1967) The role of premature beats in the initiation and the termination of supraventricular tachycardia in the Wolff–Parkinson–White syndrome. *Circulation*, **XXXVI**, 644–62.

Naumann d'Alnoncourt, C., Cardinal, R., Janse, M.J., Lüderitz, B. & Durrer, D. (1980) Effects of tocainide on ectopic impulse formation in isolated cardiac tissue. *Klinische Wochenschrift*, **58**, 227–31.

Henrick Joan Joost Wellens (born 1935)

Henrick Joan Joost Wellens was born on November 13, 1935 in Te's Gravenhaage, and is now Professor of Cardiology at Maastricht, the Netherlands. He described the mechanism of reentry tachycardia, including the Wolff–Parkinson–White syndrome. He is a pioneer of

* Lüderitz, B. (2000) History. *Journal of Interventional Cardiac Electrophysiology*, **4**, 663 (with permission).

H.J.J. Wellens, MD. Professor in Cardiology: Oration in Amsterdam
(February 25, 1974).

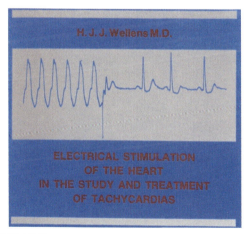

PhD thesis of H.J.J. Wellens, 1971.

clinical electrophysiology and a founder of the modern era of tachycardia management concerning atrial and ventricular arrhythmias
(Wellens et al., 1968). Wellens is a superb teacher and a master electrocardiographer. His doctoral thesis (March 18, 1971) was entitled, "Electrical
stimulation of the heart in the study and treatment of tachycardias"
(Wellens, 1971). One of the most recent interests of Wellens in arrhythmology—among many others—concerns internal cardioversion in atrial
fibrillation. He stimulated the general interest in atrial defibrillation and

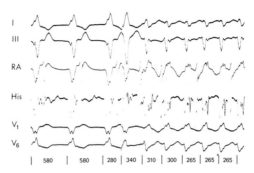

| | 580 | | 580 | | 280 | 340 | 310 | 300 | 265 | 265 | 265 |

This figure shows that clinically documented ventricular tachycardia can reproducibly be initiated and terminated by appropriately timed stimuli during programmed electrical stimulation of the heart.

in the development of an implantable atrial defibrillator. Consequently, he headed the investigators group and wrote the key paper on the implantable atrioverter for the treatment of atrial fibrillation (Wellens *et al.*, 1998). This could not have been done without his previous fundamental investigations on arrhythmic mechanisms and the effects of therapeutic interventions in human heart disease. Thus, it seems more than justified that Dr. Wellens was awarded—among many other honors and prizes—as a Pioneer in Cardiac Pacing and Electrophysiology (1995) and as Distinguished Teacher (2000) by the North American Society of Pacing and Electrophysiology. Recently, he was knighted by the Queen of the Netherlands. H.J.J. Wellens indeed provides a higher standard of excellence in the world of arrhythmology.

References
Wellens, H.J.J. (1971) *Electrical Stimulation of the Heart in the Study and Treatment of Tachycardias.* H.E. Stenfert Kroese N. V., Leiden, the Netherlands.
Wellens, H.J.J., Lau, C.P., Lüderitz, B. *et al.* for the Metrix Investigators (1998) Atrioverter: An implantable device for the treatment of atrial fibrillation. *Circulation*, **98**, 1651–6.
Wellens, H.J.J., Schuilenburg, R.M. & Durrer, D. (1968) Electrical stimulation of the heart in patients with ventricular tachycardia. *Circulation*, **46**, 216–26.

Ivan Mahaim (1897–1965)*

Ivan Mahaim was born on June 25, 1897 in Liège, Belgium. He was educated in Lausanne at the College Classique and began studying medicine

---

* Lüderitz, B. (2003) History. *Journal of Interventional Cardiac Electrophysiology*, **8**, 155 (with permission).

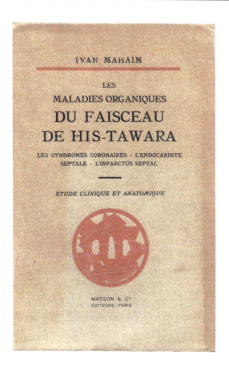

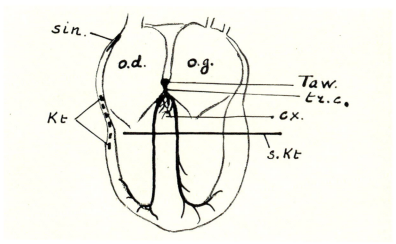

Scheme of the bundle of His-Tawara with the node of Tawara, *Taw.*, the main trunk, *tr.c.* and the Kent's fibers, *Kt.* The transverse line, *s.Kt*, represents the experimental section performed by Kent. The upper connection are represented by *cx*. Kent's fibers and the A-V paraspecific conduction through the upper connections of his bundle of His-Tawara; American Heart Journal 1947; 33: 651–653.

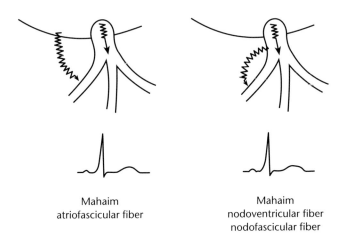

Mahaim	Mahaim
atriofascicular fiber	nodoventricular fiber
	nodofascicular fiber

in Lausanne in 1918. In 1925, he was promoted to Doctor of Medicine. His medical career started at the Hôpital Cantonal in Lausanne and at the Lausanne Policlinique, and continued at the Institute of Pathology. He was a fellow of Professor Wenckebach in Vienna, Austria (1926) and of Professor Clerc in Paris, France (1927). In 1929, he became an Associate Professor at the University of Lausanne. Mahaim wrote 100 papers that were published in all of the important journals of his time. His most influential works were his publications on histologic research concerning the connections of the bundle of His, which were a resounding success in Europe in 1937 as it provided the basis for later electrophysiological discoveries. Another of his publications, which is used as a reference today, is his book *Les Maladies organiques du faisceau de His—Tawara* (Mahaim, 1931), which was devoted to the tumors of the heart, providing a detailed analysis of more than 400 cases. This book anticipates the possibility of cardiac surgery, which at that time was considered science fiction. His last big *oeuvre* was devoted to Beethoven and was published in 1964. It is still a reference for musicians as it is the most detailed analysis and description of the background on Beethoven's last quartuors. With these fine works, Mahaim has proved his outstanding personality as a detailed analyst, a great artist, and a devoted writer. In-depth study of the work of Mahaim will certainly lead one to an outstanding human being, to a pioneering time of medicine and cardiology, and to the harmony between the first concepts of electrophysiology and the quartuors of Beethoven.

References
Lüderitz, B. (2002) *History of the Disorders of Cardiac Rhythm*, 3rd edn. Futura, Armonk, NY.

Mahaim, I. (1931) *Les Maladies organiques du faisceau de His—Tawara. Etude clinique et anatomique.* Masson & Cie., Paris.
Steinbeck, G. (1996) Rhythmusstörungen des Herzens. In: *Klinische Kardiologie.* (eds E. Erdmann, G. Riecker), pp. 582–645. Springer, Heidelberg.

Paul M. Zoll (1911–1999)*

Dr. Zoll was born and educated in Boston, Massachusetts. He attended Harvard Medical School and trained at the Boston Beth Israel Hospital, in Boston, where he remained for the duration of his career. In an epochal publication in 1952, he described cardiac resuscitation via electrodes on the bare chest with 2-ms-duration pulses of 100–150 volts across the chest, at some 60 stimuli per minute. This initial clinical description launched widespread evaluation of pacing and the recognition by the medical profession and the public that the asystolic heart could be stimulated to beat, and was the basis for many developments to come. In 1956, Zoll published a description of a transcutaneous approach that terminated ventricular fibrillation with a much larger shock. He later described a similar termination of ventricular tachycardia. He was

* Lüderitz, B. (2002) History. *Journal of Interventional Cardiac Electrophysiology,* **7,** 193 (with permission).

The New England

Journal of Medicine

Copyright, 1956, by the Massachusetts Medical Society

| Volume 254 | MARCH 22, 1956 | Number 12 |

TREATMENT OF UNEXPECTED CARDIAC ARREST BY EXTERNAL ELECTRIC STIMULATION OF THE HEART*

Paul M. Zoll, M.D.,† Arthur J. Linenthal, M.D.,‡ Leona R. Norman, M.D.,§
Milton H. Paul, M.D.,¶ and William Gibson, M.D.‖

BOSTON

CARDIAC arrest may occur unexpectedly during various diagnostic and therapeutic procedures, particularly under anesthesia. Though infrequent (1 in every 500 to 5000 operations[1,2]), each accident or four minutes if cardiac and cerebral function is to return unimpaired. In view of this desperate urgency, the recognized risk of occasional unnecessary thoracotomy for conditions other than cardiac

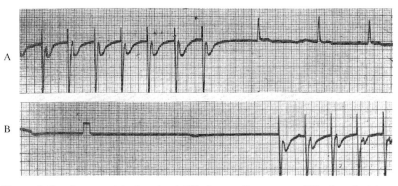

Resuscitation from ventricular standstill after cardiac surgery. Effective electric stimulation during ventricular standstill is followed by spontaneous cardiac beats (A). An episode of ventricular standstill is terminated by effective external stimulation (B).

awarded the Lasker Award in 1973, and was recognized as a Pioneer in Cardiac Pacing by the North American Society of Pacing and Electrophysiology in 1989.

References

Lüderitz, B. (2002) *History of the Disorders of Cardiac Rhythm*, 3rd edn. Futura, Armonk, NY.

Zoll, P.M. (1952) Resuscitation of the heart in ventricular standstill by external electric stimulation. *New England Journal of Medicine*, **247**, 768–71.

Zoll, P.M., Linenthal, A.J., Norman, L.R., Paul, M.H. & Gibson, W. (1956) Treatment of unexpected cardiac arrest by external electric stimulation of the heart. *New England Journal of Medicine*, **254**, 541–6.

R. Elmqvist/Å. Senning; Forty-five years of implantable cardiac pacemakers*

In October 1958 a new chapter in the history of medicine was started. In that year a doctor in Sweden, Professor Åke Senning, who was then an assistant of the well-known heart surgeon Clarence Crafoord, and subsequently became a famous surgeon in Switzerland, succeeded for the first time in treating a patient who was suffering from an anomaly of heart rhythm with the operative implantation of a pacemaker. Professor Åke Senning had been communicating with Dr. Rune Elmqvist. After

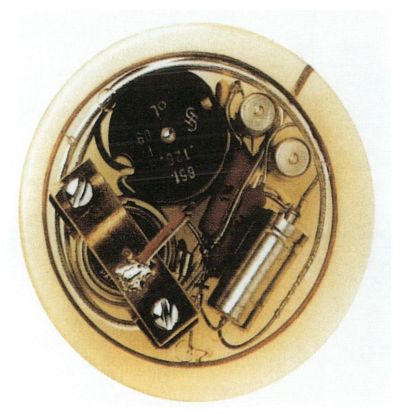

The first implantable cardiac pacemaker. Note the large battery, two old-style transistors, and condensers.

---

* Lüderitz, B. (1998) History. *Journal of Interventional Cardiac Electrophysiology*, **2**, 221 (with permission).

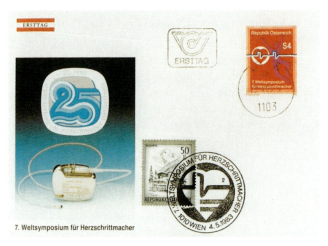

7. Weltsymposium für Herzschrittmacher

Austrian first-day issue, released on the occasion of the Seventh World Symposium on Cardiac Pacing (Vienna, May 1–5, 1983) commemorating 25 years of cardiac pacing with implantable pacemakers.

intensive research work, they succeeded in constructing a device that was so small it could be implanted under the skin but with battery capacity sufficient for long-term use.

A pacemaker consists of an electronic part and a battery. The demands upon the quality of this apparatus are to some extent much higher than for any aircraft, missile, or satellite. The electronic part is the brain of the pacemaker. It works like a tiny computer. The pacemaker does not only give electric pulses to the heart, it can also feel when one's heart can work without any help. It only starts stimulating it again when help is needed.

Millions of people suffering from different disturbances of the electric nervous system have received pacemakers. Whereas the first operations demanded huge surgical effort, nowadays a simple operation, which is performed every day in thousands of hospitals all over the world, is sufficient. Over the years, pacemakers have achieved a high-quality standard and, as every pacemaker is also checked very thoroughly, failure in its electronics is very rare nowadays.

Reference
Elmqvist, R. & Senning, Å. (1960) An implantable pacemaker for the heart. In: *Medical Electronics, Proceedings of the Second International Conference on Medical Electronics, Paris 1959.* (ed. C.N. Smyth). Iliffe & Sons, London.

F. Dessertenne; Torsade de pointes*

Torsade de pointes is a specific form of dangerous ventricular tachycardia in which an undulating series of ventricular beats appear on the QRS axis. These may be considered a specific form of ventricular flutter wave. Dessertenne first described these torsades de pointes waves in 1996 when he observed this rhythm disorder in an 80-year-old female patient with complete intermittent atrioventricular (AV) block. In this case, the cause of her recurring syncopal episodes was torsade de pointes tachycardia rather than third-degree AV block or bradyarrhythmias, as was originally suspected.

La tachycardie ventriculaire
à deux foyers opposés variables

Par F. DESSERTENNE (*)

L'étude que nous avons faite d'un certain nombre de tracés de fibrillation ventriculaire recueillis dans le service de réanimation de l'hôpital Lariboisière comportait une description et une hypothèse.

Pour le cardiologue habitué à reconnaître des ventriculogrammes et à ne rencontrer que des variations brusques de l'amplitude du tracé, lors de l'extra-systole par exemple, la description mettait l'accent sur la succession ininterrompue d'oscillations irrégulières présentant dans l'ensemble des variations progressives d'amplitude autour d'une ligne de référence, en fuseaux.

L'hypothèse était qu'un tel aspect évoque un phénomène de battement produit par les combinaisons de l'activité électrique de plusieurs centres, tantôt en phase et tantôt en opposition de phase.

Or, il s'en faut que toutes les variations progressives d'amplitude d'un tracé relèvent de la fibrillation des ventricules.

Nous en avons rencontré au cours des accidents syncopaux du bloc complet du faisceau de His, dont l'aspect en torsades de pointes paraît relever d'une tachycardie ventriculaire à deux foyers opposés variables.

C'est une observation clinique récente qui nous a mis sur cette voie.

---

* Lüderitz, B. (1997) History. *Journal of Interventional Cardiac Electrophysiology*, **1**, 327 (with permission).

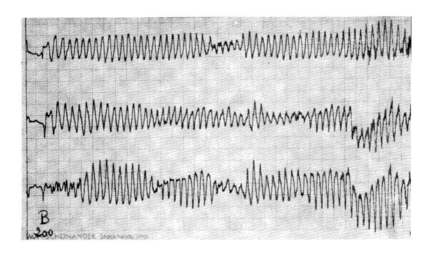

References
Dessertenne, F. (1966) La tachycardie ventriculaire à deux foyers opposés variables. *Archives des maladies du coeur et des vaisseaux*, **59**, 263–72.
Lüderitz, B. (1993) *History of the Disorders of Cardiac Rhythm*. Futura, Armonk, NY.

Jervell and Lange-Nielsen syndrome: Anton Jervell, Fred Lange-Nielsen

Jervell and Lange-Nielsen reported on a case of hereditary, functional syncopal arrhythmia in combination with profound congenital deafness in a Norwegian family with six children.

The parents and two children were healthy, but four of the children were deaf–mutes. All of these four children suffered from episodes of loss of consciousness and exhibited a long QT interval on the electrocardiogram (ECG). Three of them died suddenly.

Jervell, A. and Lange-Nielson, F. *Congenital Deaf–Mutism, Functional Heart Disease with Prolongation of the QT Interval, and Sudden Death.*

CONGENITAL DEAF-MUTISM, FUNCTIONAL HEART DISEASE
WITH PROLONGATION OF THE Q-T INTERVAL,
AND SUDDEN DEATH

Anton Jervell, M.D., and Fred Lange-Nielsen, M.D.

Tönsberg, Norway

A COMBINATION of deaf-mutism and a peculiar heart disease has been observed in 4 children in a family of 6. The parents were not related, and were, as the other 2 children, quite healthy and had normal hearing.

The deaf-mute children, who otherwise seemed quite healthy, suffered from "fainting attacks" occurring from the age of 3 to 5 years. By clinical and roentgen examination, which was performed in 3 of the children, no signs of heart disease could be discovered. The electrocardiograms, however, revealed a pronounced prolongation of the Q-T interval in all cases.

Three of the deaf-mute children died suddenly at the ages of 4, 5, and 9 years, respectively.

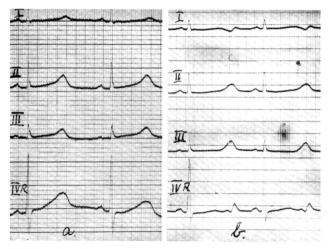

Tormond J. (a) ECG July 20, 1953, during rest. Leads I, II, III, IV R. QT = 0.50 second RR = 0.88 second (b) ECG July 20, 1953, after stair running Leads I, II, III, IV R QT = 0.60 second RR = 0.86 second.

Romano–Ward syndrome: Cesarino Romano (born 1924), Owen Conor Ward (born 1923)*

The Romano–Ward syndrome is a functional, hereditary syncopal cardiac disorder characterized by a prolongation of the QT interval. This

---

* Lüderitz, B. (1998) History. *Journal of Interventional Cardiac Electrophysiology*, **2**, 101 (with permission).

disease is a variation of the long QT syndrome in deaf patients first described in 1957 by Anton Jervell and Fred Lange-Nielsen in their work *Congenital Deaf–Mutism, Functional Heart Disease with Prolongation of the QT Interval, and Sudden Death* (Am Heart J 1957; 54: 59–68).

Reference
Lüderitz, B. (1993) *History of the Disorders of Cardiac Rhythm*. Futura, Armonk, NY.

Max Gustav Julius Schaldach (1936–2001)*

On May 5, 2001, Max Schaldach, the physicist and entrepreneur, died after his plane crashed in the vicinity of Nuremberg, Germany. He was 64 years old.

Prof. Dr. Max Gustav Julius Schaldach was born in Berlin, Germany, on July 19, 1936, the son of a Pommeranian family. He was Professor and Chairman of Biomedical Engineering at Friedrich–Alexander University, Erlangen—Nuremberg, Professor of Postgraduate Studies at the State School of Medicine, São José do Rio Preto in Brazil, and Professor of

---

* Lüderitz, B. (2001) History. *Journal of Interventional Cardiac Electrophysiology*, **5**, 515–16 (with permission).

Biophysics at Lomonosov University in Moscow. In addition, he was the founder and owner of several high-technology companies, including Biotronik, whose name is associated worldwide with the production of widely used cardiac pacemakers and defibrillators. In 1963, the company developed the first German pacemaker and started marketing the devices, only 5 years after the initial implantation of a pacemaker prototype in Sweden. Today, 850 full-time employees work at the Biotronik headquarters in Berlin—Neukölln, and affiliated Biotronik companies employ 2300 people worldwide.

Upon completion of his college studies, Max Schaldach received a grant from the Heinrich Hertz Foundation. He completed a BSc degree in Physics in 1961, and a MSc degree in Physics in 1964 in Berlin. In 1966, he earned his PhD in Engineering, and in 1968 he completed his professional dissertation at the Technical University-Berlin with his thesis, "Potential- and charge-distribution in the solid-electrolyte interface." In 1970, he became a Full Professor at Friedrich–Alexander University, Erlangen–Nuremberg.

In its broadest sense, his main focus was electrophysiology. He specialized in electrotherapy of the heart, electronic implants, cardiac pacemakers, neurostimulators, defibrillators, artificial organs, circulation-assist devices, and interventional cardiology. These, together with his other areas of research, led to more than 100 patents.

The Biotronik company, under Schaldach's personal leadership, was particularly active in Western Europe, as well as in North and South America. He fostered intensive academic and business relationships with Brazil, Israel, and the former Eastern Block countries. Because of his special relationship with Russia, in 1998 Schaldach became a member of the Russian Academy of Natural Sciences in the Mathematics Division.

Among his many distinguished memberships in scientific societies, academies, and other organizations, his affiliation with the Society of German Physical Scientists and Physicians is particularly noteworthy.

He graciously accepted many awards and honors. His most prestigious recognitions included: 1989, Most Distinguished Cross of Merit from the Federal Republic of Germany for Outstanding Technical and Industrial Developments; 1992, Honorary Doctorate from the Medical Academy, Kaunas, Lithuania; 1994, Honorary Doctorate, Lomonosov University, Moscow; 1994, Honorary Member of the German Society for Biomedical Engineering; 1999, Honorary Doctorate from the Russian Medical Academy, St. Petersburg. In the year 2000, he became an Honorary Citizen of the City of Erlangen. Furthermore, he was the recipient of the "Year 2000 Career Achievement Award" from the Engineering in Medicine and Biology Society (EMBS) of the Institute of Electrical and Electronic Engineers (IEEE). He received this award for his, "contributions to the application of engineering and technology to the fields of medicine and biology, as well as his global leadership for the benefit of humanity by disseminating knowledge, setting standards, fostering professional development, and recognizing excellence."

Schaldach is the author or the coauthor of over 1000 scientific publications. He supervised more than 20 professional dissertations and 120 master and doctoral theses during his scientific teaching career.

His personal interests included history and philosophy, ancient languages, and classical music. He had a life-long passion for flying, which he enjoyed to the tragic end of his life.

Schaldach was a visionary entrepreneur who was not seeking acclaim. He was a generous humanitarian who was thoughtful towards others. He avoided publicity and continually concentrated his efforts on people who were close to him.

Schaldach's motto, to which he dedicated his life and his companies, was, "Technology helping to heal." Schaldach is survived by two sons (Max and Marcus) and a daughter (Brita). The company will be continued under the leadership of his elder son, Dr. Max Schaldach, who studied materials science and received his doctorate for his thesis in microcircuit hybrid technology for implants. His many years of management experience in another internationally renowned company will be advantageous to him in his new responsibilities.

With the passing of Max Schaldach, the physical sciences and biomedical engineering have lost one of their truly great visionary pioneers. In a singular fashion, he governed the osmosis between innovative research and entrepreneurship, which led to application-oriented and successful "spin-off" companies.

Jacques Edmond Mugica (1933–2002)*

Jacques Mugica was born in southern France on February 28, 1933. His mother was a first-generation Russian physician. Before attending medical school, Mugica pursued his interest in archeology. Between 1959 and 1964, he joined the French army and participated in the Algerian war.

Dr. Mugica graduated from the University of Paris in 1965, and subsequently received training in cardiopulmonary medicine at the Hôpital Lariboisière in the Department of Prof. Yves Bouvrain (his father-in-law), who coinvented cardiac resuscitation and the pacemaker/defibrillator

---

* Lüderitz, B. (2003) History. *Journal of Interventional Cardiac Electrophysiology*, **8**, 227–8 (with permission).

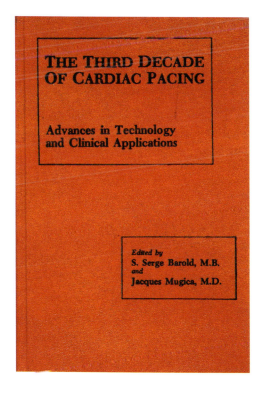

combination. Mugica then joined the Val d'Or Medical Center in St. Cloud, Paris, where, in the early years, he practiced pulmonary medicine, cardiology, and intensive care. In 1968, Dr. Mugica established the first independent, multidisciplinary, private pacemaker center in the world. Starting with 60 pacemaker implantations a year, the center grew rapidly to almost 1000 implantations in 1994. The Val d'Or Center quickly became an important center for advanced health care, teaching, research, and postgraduate training for many international students.

Dr. Mugica lectured all over the world and was a leading authority on pacemaker leads. He pioneered a number of important developments in pacing, including a pacemaker with automatic threshold detection, screw-in leads, Holter systems in pacemakers, and the concept of the fully automatic pacemaker, now widely recognized as an achievable goal. In a collaborative effort with his coworkers, particularly Dr. Serge Cazeau, he introduced the first four-chamber pacing system, with a view to resynchronizing the heart as means of managing congestive heart failure; he published these results in *Pacing and Clinical Electrophysiology: PACE* in November 1994. Dr. Mugica served on the editorial boards of five journals, including *Pacing and Clinical Electrophysiology: PACE*. He published over 250 articles and authored or coauthored a total of ten books. Particularly noteworthy is his reference book *The Third Decade of Cardiac Pacing. Advances in Technology and Clinical Applications*, which he coedited with S. Serge Barold in 1982 (Barold & Mugica, 1982).

Dr. Mugica created the well-known Cardiostim organization when he realized that progress in cardiac pacing and the proper application of new technology require a close collaboration between physicians and industry. The first Cardiostim symposium was held in 1978. It soon became one of the most important meetings in pacing and electrophysiology, with nearly 5000 participants from all over the world in 2002.

Dr. Mugica, Chief Physician of the Cardiac Pacing Department St. Cloud and Chairman of the Cardiostim Association, received many awards for his contributions to cardiology. These honors include the Medal of the City of Paris, the 1995 Distinguished Service Award of the North American Society for Pacing and Electrophysiology, and many others.

Jacques Edmond Mugica, a pioneer in cardiac pacing and the founder of Cardiostim, the electrophysiology and cardiac rhythm disorders organization, passed the organization's presidential baton to Dr. Philippe Ritter at Cardiostim 2002. There was already a feeling of sadness when he retired from his most cherished activity. At the age of 69, Dr. Mugica died on December 12, 2002, but his legacy continues.

References
Barold, S.S. & Mugica, J. (eds) (1982) *The Third Decade of Cardiac Pacing. Advances in Technology and Clinical Applications*. Futura, Mount Kisco, NY.
Bouvrain, Y. (1996) *En auscultant les cœurs. Soixante ans de cardiologie à Paris*. Frison–Roche, Paris.

Cazeau, S., Ritter, P., Bakdach, S. *et al.* (1994) Four-chamber pacing in dilated cardiomyopathy. *Pacing and Clinical Electrophysiology: PACE,* **17**(II), 1974–9.

Lüderitz, B. (2002) *History of the Disorders of Cardiac Rhythm,* 3rd edn. Futura, Armonk, NY.

Helmut Weber; Catheter technique for closed-chest ablation of an accessory atrioventricular pathway*

The New England
Journal of Medicine
©Copyright, 1983, by the Massachusetts Medical Society

| Volume 308 | MARCH 17, 1983 | Number 11 |

| 3400 Göttingen,
Federal Republic of Germany | HELMUT WEBER, M.D.
LOTHAR SCHMITZ, M.D.
University of Göttingen |

CATHETER TECHNIQUE FOR CLOSED-CHEST
ABLATION OF AN ACCESSORY
ATRIOVENTRICULAR PATHWAY

To the Editor: In drug-resistant arrhythmias caused by circuit movements through accessory atrioventricular pathways, local cryotherapy appears to be the most successful treatment.[1] However, this method requires open-heart surgery with its attendant risk. A catheter technique for closed-chest ablation of the normal atrioventricular conduction system has been reported in the *Journal.*[2] The indication for this procedure is limited by the subsequent need for permanent pacing in all such patients. Furthermore, life-threatening arrhythmias due to rapid atrioventricular conduction through the accessory pathway in cases of atrial flutter or fibrillation cannot be excluded, and the initiation of atrial arrhythmias by retrograde

---

* Lüderitz, B. (2000) History. *Journal of Interventional Cardiac Electrophysiology,* **4,** 327 (with permission).

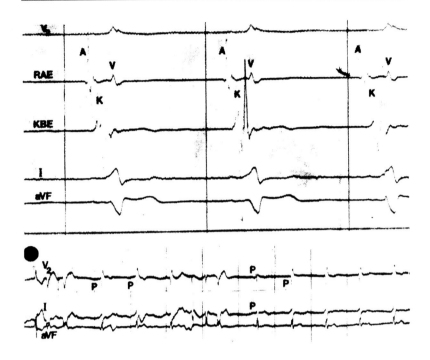

References

Gallagher, J.J., Sealy, W.C., Anderson, R.W. *et al.* (1977) Cryosurgical ablation of accessory atrioventricular connections: A method for correction of the preexcitation syndrome. *Circulation*, **55**, 471–9.

Gallagher, J.J., Svenson, R.H., Kasell, J.H. *et al.* (1982) Catheter technique for closed-chest ablation of the atrioventricular conduction system: A therapeutic alternative for the treatment of refractory supraventricular tachycardia. *New England Journal of Medicine*, **306**, 194–200.

Weber, H. & Schmitz, L. (1983) Catheter technique for closed-chest ablation of an accessory atrioventricular pathway. *New England Journal of Medicine*, **308**, 653–4.

Seymour Furman (born 1931)*

Dr. Seymour Furman was born in New York City in 1931, graduated from Washington Square College, New York University, and, in 1955, received his MD degree from the State University of New York, College of Medicine Downstate Medical Center. He was the first to show the feasibility of non-surgical treatment of heart disease from inside the

---

* Lüderitz, B. (2002) History. *Journal of Interventional Cardiac Electrophysiology*, **7**, 113 (with permission).

AN INTRACARDIAC PACEMAKER FOR STOKES–ADAMS SEIZURES*

Seymour Furman, M.D.,† and John B. Schwedel, M.D.‡

NEW YORK CITY

THIS paper is a report of the clinical application of an intracardiac pacemaker in the therapy of severe or intractable Stokes–Adams seizures associated with heart block. This is the first recorded description of such clinical use.

The resuscitation of the heart in asystole has been a goal sought for many years. It has been known for a considerable time that the mammalian heart, either as the intact organ or as a strip of tissue, is and Watkins.[4] There, electric impulses were conveyed to the myocardium through the use of electrodes embedded directly in the myocardium itself. Lately, these electrodes have come into general use for heart block after repair of congenital cardiac defects[5,6] Stimulation of the heart from without the chest and from an electrode buried in the myocardium has been proved to be effective and safe for the maintenance of the heart beat. Both technics have major

heart. On July 16, 1958 at Montefiore Hospital in the Bronx, New York City, Dr. Furman paced the right ventricle with a transvenous electrode catheter in a patient with complete atrioventricular block. This accomplishment defined the first step in the development of clinical cardiac electrophysiology as we know it today.

For the next 43 years, Dr. Furman has remained a world leader in electrical stimulation of the heart. His contributions are so numerous that many are now taken for granted and their origin often forgotten. They include the establishment of factors controlling electrode efficiency by utilizing the strength–duration curve, increasing battery longevity by reducing current drain, and developing the concept of organized device follow-up in a pacemaker clinic and transtelephonic monitoring.

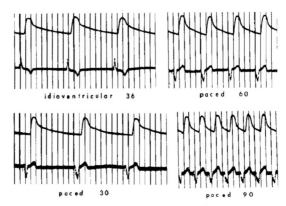

idioventricular 36   paced 60

paced 30   paced 90

Idioventricular rhythm and varied imposed rates, with accompanying femoral-artery tracings (S. Furman, J.B. Schwedel, 1959).

Dr. Furman was the Secretary General of the Second World Symposium on Cardiac Pacing in 1967. He cofounded the North American Society of Pacing and Electrophysiology (NASPE) and has remained an active member, especially as the leader of the oral history project. Dr. Furman has been Editor of *Pacing and Clinical Electrophysiology: PACE* since its inception in 1978. He also piloted an examination process (NASPExAM) to establish intellectual standards in cardiac pacing and defibrillation. Dr. Furman has been President of both NASPE (in 1980) and NASPExAM (1985–1999). The recipient of many awards from prestigious learned societies, Dr. Furman continues his active professional life as Professor of Medicine and Surgery at Albert Einstein College of Medicine in New York.

References
Furman, S. & Schwedel, J.B. (1959) An intracardiac pacemaker for Stokes–Adams seizures. *New England Journal of Medicine*, **261**, 943–8.
Lüderitz, B. (2002) *History of the Disorders of Cardiac Rhythm*, 3rd edn. Futura, Armonk, NY.

John A. McWilliam (1857–1937)—"Pacemaker syndrome," 70 years before the first pacemaker was implanted*

Around 100 years ago, in Aberdeen, Scotland, John A. McWilliam documented the effect of electric current on the contraction of the isolated

* Lüderitz, B. (2001) History. *Journal of Interventional Cardiac Electrophysiology*, **5**, 341 (with permission).

ELECTRICAL STIMULATION OF THE HEART
IN MAN.

By JOHN A. McWILLIAM, M.D.,

Professor of the Institutes of Medicine in the University of Aberdeen.
(From the Physiological Laboratory of the University of Aberdeen.)

IT is, of course, only in a very limited number of the cases of
cardiac failure that the question of artificial excitation of the
heart beat becomes one of practical importance. In the majority
of instances where a more or less sudden heart stoppage occurs
there are underlying conditions which obviously render direct

BRITISH MEDICAL JOURNAL.

BEING THE JOURNAL OF THE BRITISH MEDICAL ASSOCIATION.

EDITED FOR THE ASSOCIATION BY ERNEST HART.

LONDON: SATURDAY, JANUARY 5, 1889.

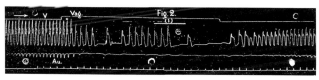

(To be read from left to right). Cat's heart. Uppermost tracing shows the period
of vagus stimulation. Second tracing (marked V) records the action of the
ventricles (upward movement = contraction); and the third tracing (marked Au)
the action of the auricles (downward movement = contraction). The lowest line
indicates half seconds. After the heart had been depressed and enfeebled by
vagus stimulation a periodic series of (eight) induction shocks was applied to the
ventricles. The resulting group of beats is marked (1). The individual beats are
much improved in strength as compared with the spontaneous beats occurring
before and after. The beneficial effect of direct excitation is very apparent.

heart of a cat. He concluded that it had to be possible to treat bradycardia
and cardiac arrest with electrical stimulation in humans too. McWilliam
realized that the heart could be stimulated to adhere to a certain frequency by regular impulses which were controlled by a metronome, and
that a decrease in blood pressure as caused by bradycardia and cardiac
arrest may be reverted after normalizing the heart rate.

McWilliam, who performed his investigations without the help of the
electrocardiogram (ECG), described a decrease in stroke volume as an
effect of *depressed atrial contraction due to stimulation of the left ventricular
apex*. He concluded that simultaneous atrial and ventricular stimulation
by means of a pacemaker would be hemodynamically advantageous.

McWilliam thus represents a distinguished figure in a movement researching the manipulation of the heart beat via electric stimulation.

It has been known since Galvani's experiments that electrically stimulated muscles contract. Thus, it is not an accident that Galvani's nephew G. Aldini used galvanic energy to resuscitate the motionless heart. The fact that the heart may be stimulated by means of electricity had already been described by Burns 10 years earlier.

References
Barold, S.S. Personal communication, 2001.
De Silva, R. (1989) John MacWilliam, evolutionary biology and sudden cardiac death. *Journal of the American College of Cardiology,* **14,** 1843–9.
McWilliam, J.A. (1889) Electrical stimulation of the heart in man. *British Medical Journal,* **1,** 348–50.

Philippe Coumel (1935–2004); Early reports of multisite pacing for artificial preexcitation of the arrhythmia substrate*

In 1967 Coumel's group from Paris, France, used programmed atrial and ventricular stimulation to unravel the mechanism and diagnosis of per-

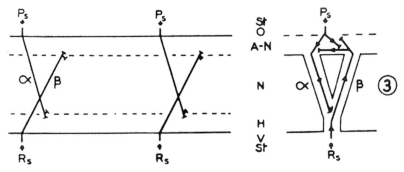

Coumel *et al.* (1967) indicated in a three-part diagram the three ways that pacing could prevent permanent junctional reciprocating tachycardia. The third panel, reproduced here, shows the mechanism they believed prevented the tachycardia in their patient.

> *Stimulation auriculaire et ventriculaire simultanée; l'envahissement simultané de la voie α par l'influx d'origine ventriculaire et de la voie β par l'influx d'origine ventriculaire fait qu'aucun de ces deux influx ne peut envahir la voie opposée, et que le rhythme réciproque ne peut réapparaître.*

* Lüderitz, B. (2001) History. *Journal of Interventional Cardiac Electrophysiology,* **5,** 511 (with permission).

manent junctional reciprocating tachycardia in a drug-refractory patient. Their elegant studies showed that the tachycardia was sustained by a reentrant or reciprocal rhythm. Coumel *et al.* (1967) postulated that the process involved dual atrioventricular (AV) nodal pathways with a slow conducting retrograde pathway. The latter was shown many years later to be an accessory pathway with the same electrophysiologic properties originally proposed by Coumel *et al.* (1967). A permanent bipolar pacemaker (designed for single chamber pacing) was implanted with the cathode on the anterior surface of the right ventricle and the anode on the interatrial groove. Simultaneous atrial and ventricular pacing at a rate of 100 beats per minute was well tolerated and caused disappearance of the tachycardia. Their pacing arrangement is similar to the split bipolar (sometimes called dual unipolar) configuration used in the last decade for multisite pacing of one electrical chamber, either the atrium or ventricle from a single pacemaker port or output. Coumel's group in fact, used a pacemaker to change the electrophysiologic substrate by producing preexcitation, an approach conceptually similar to the present practice of multisite pacing for the prevention of atrial tachyarrhythmias. Accordingly, Coumel's group was the first to use multisite pacing for the treatment of a cardiac arrhythmia. Furthermore their report antedated that of Ryan *et al.* (1968), widely considered to represent the first implanted antitachycardia pacemaker. (See page 211).

References

Coumel, Ph., Cabrol, C., Fabiato, A. *et al.* (1967) Tachycardie permanente par rhythme réciproque. I. Preuve du diagnostic par stimulation auriculaire et ventriculaire. II. Traitement par l'implantation intracorporelle d'un stimulateur cardiaque avec entrainement simultané de l'oreillette et du ventricule. *Archives des maladies du coeur et des vaisseaux*, **12**, 1830–64.

Daubert, C., Berder, V., Gras, D. *et al.* (1994) Atrial tachyarrhythmias associated with high degree interatrial conduction block: Prevention by permanent atrial resynchronization. *European Journal for Cardiac Pacing and Electrophysiology*, **1**, 35–44.

Daubert, C., Mabo, P., Berder, V. *et al.* (1990) Arrhythmia prevention by permanent atrial resynchronization in patients with advanced interatrial block (Abstract). *European Heart Journal*, **11**, 237.

Delfaut, P., Saksena, S., Prakash, A. *et al.* (1998) Long-term outcome in patients with drug-refractory atrial flutter and fibrillation after single- and dual-site right atrial pacing for arrhythmia prevention. *Journal of the American College of Cardiology*, **32**, 1900–8.

Ryan, G.F., Easley, R.M., Zaroff, L.I. *et al.* (1968) Paradoxical use of a demand pacemaker in the treatment of supraventricular tachycardia due to the Wolff–Parkinson–White syndrome. Observation on the termination of reciprocal rhythm. *Circulation*, **38**, 1037–43.

Andreas Roland Grüntzig (1939–1985)*

Andreas Roland Grüntzig was born on June 25, 1939 in Dresden, Germany. As internist in the Kantonsspital of Zurich, Switzerland, he introduced percutaneous transluminal coronary angioplasty to human medicine. He modified a technique innovated by Dotter and Judkins in 1964, who tried to mechanically dilate short stenoses and occlusions of leg arteries with a coaxial double catheter of 8 French (2.6 mm) and 12 French (3.6 mm).

In 1973, Porstmann presented a catheter with a latex balloon at the tip but caved by a corset, which was the longitudinally sliced Teflon catheter. In 1976, Grüntzig described in the "Klinische Wochenschrift" a short sausage-shaped polyvinyl chloride balloon located at the tip of a double lumen catheter. After several animal experiments, postmortem examinations, and interoperative dilatations, he decided to use the technique for coronary dilatation in humans. The ideal candidate was selected in September 1977. Both the doctor and his first patient were 38

* Lüderitz, B. (2004) History. *Journal of Interventional Cardiac Electrophysiology*, **10**, 177–8 (with permission).

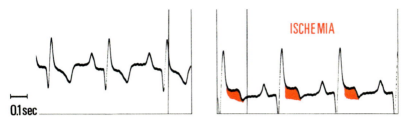

Original electrocardiogram (ECG) registration of A.R. Grüntzig's early dog experiment. Before and during occlusion of the circumflex branch of the left coronary artery causing ischemia (ST elevation). Grüntzig A. Die perkutane transluminale Rekanalisation chronischer Arterienverschlüsse mit einer neuen Dilatationstechnik. Witzstrock, Baden-Baden, 1997.

Klin, Wschr. 54, 543-545 (1976)

**Perkutane Dilatation von Coronarstenosen —
Beschreibung eines neuen Kathetersystems**

A. Grüntzig
Departement für Innere Medizin (Proff. P. Frick, A. Labhart, W.
Siegenthaler) und dem Röntgendiagnostischen Zentralinstitut
(Prof. J. Wellauer) der Universität Zürich

**Percutaneous Dilatation of Experimental Coronary
Artery Stenosis—Description of a new Catheter
System**

Summary. A technical description of a catheter system is given for percutaneous dilatatio0n of experimental stenosis in small arteries (e.g. left coronary artery). The system consists of two catheters: the first is positioned into the orifice of the small artery and guides the second, double-lumen-dilatation catheter into the artery. During the manipulation ischemia in the peripheral area is prevented by contineous perfusion with oxygenated blood taken from the femoral artery by a roller-pump. The catheter system was tested in canine experiments.

Key words: Dilatation catheter—Coronary artery stenosis—Percutaneous dilatation—Myocardial perfusion.

---

years of age. The patient had severe effort angina related to a single LAD lesion. He enthusiastically gave his consent and the procedure was performed on September 16, 1977. In 1984, Andreas Grüntzig recorded his personal account of the first case: "The dilatation catheter was advanced to the stenosis with no difficulty. . . . To the surprise of all of us, no ST-elevation, ventricular fibrillation or even extrasystole occurred and the patient had no chest pain. . . . Two inflations were performed. . . . After the balloon deflation, the distal coronary pressure normalized as compared to the aortic pressure. Everyone was surprised about the ease of the procedure . . ."

Coronary angioplasty was born and several major improvements and new devices were to follow: steerable guidewire (J. Simpson in 1982); long technique guidewire (M. Kaltenbach in 1984); monorail technique (T. Bonzel in 1986); directional atherectomy (first case by J. Simpson in 1985); stent implantation (first case by J. Puel in March 1986); rotary ablation (first case by M. Bertrand in January 1988).

Today millions of procedures have been performed, and a new discipline "interventional cardiology" has been created, which extends well beyond coronary arteries—to valvular diseases, congenital heart diseases, and particularly cardiac arrhythmias, etc. Andreas R. Grüntzig and his wife Margaret Ann Thornton Grüntzig died in a plane crash in Monroe County, Georgia, on October 27, 1985.

References

Dotter, C.T. & Judkins, M.P. (1964) Transluminal treatment of arteriosclerotic obstruction: Description of a new technique and a preliminary report of its application. *Circulation*, 3, 654–70.

Grüntzig, A. (1976) Perkutane Dilatation von Coronarstenosen—Beschreibung eines neuen Kathetersystems. *Klinische Wochenschrift*, 54, 543–5.

Grüntzig, A. (1978) Transluminal dilatation of coronary-artery stenosis (letter). *Lancet*, 1, 263.

Grüntzig, A.R., Senning, Å. & Siegenthaler, W.E. (1979) Non-operative dilatation of coronary–artery stenosis: Percutaneous transluminal coronary angioplasty. *New England Journal of Medicine*, 301, 61–8.

Lüderitz, B. (2002) *History of the Disorders of Cardiac Rhythm*, 3rd edn. Futura, Armonk, NY.

Porstmann, W. (1973) Ein neuer Korsett-Ballonkatheter zur transluminalen Rekanalisation nach Dotter unter besonderer Berücksichtigung von Obliterationen an den Beckenarterien. *Radiol Diagn*, 14, 239–44.

Mauricio B. Rosenbaum (1921–2003)*

Mauricio B. Rosenbaum was born in Cordoba, Argentina, on August 25, 1921. In 1951, he obtained his MD degree from the University of Cordoba and then trained in internal medicine at the National Hospital in the same city. In 1952, he moved to Buenos Aires, where he trained in cardiology at the Ramos Mejia Hospital until 1954. His specific fields of investigation and expertise include epidemiology, diagnostic and therapeutic

---

* Lüderitz, B. (2002) History. *Journal of Interventional Cardiac Electrophysiology*, 6, 287 (with permission).

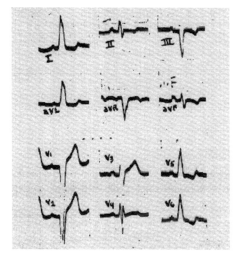

Left-bundle-branch-block patterns. From the *American Heart Journal*, 1955; 50: 38.

aspects of Chagasic cardiomyopathy, electrophysiological and therapeutic effects of antiarrhythmic drugs in life-threatening cardiac arrhythmias, and intermittent atrioventricular conduction disturbances. His
name is directly linked to the concept of "hemiblock." Moreover, his latest

discovery, the heart's long-lasting memory reflected in the T-wave of the electrocardiogram (ECG), remains a fascinating puzzle in modern electrocardiography.

References

Lüderitz, B. (2002) *History of the Disorders of Cardiac Rhythm*, 3rd edn. Futura, Armonk, NY.

Rosenbaum, M.B. (1964) Chagasic myocardiopathy. *Progress in Cardiovascular Diseases*, **7**, 199.

Rosenbaum, M.B., Blanco, H.H., Elizari, M.V., Lazzari, J.O. & Davidenko, J.M. (1982) Electrotonic modulation of the T-wave and cardiac memory. *The American Journal of Cardiology*, **50**, 213.

Rosenbaum, M.B., Elizari, M.V. & Lazzari, J.O. (1967) *Los Hemibloqueous*. Paidos, Buenos Aires.

Rosenbaum, M.B. & Lepeschkin, E. (1955) Bilateral bundle branch block. *American Heart Journal*, **50**, 38.

Agustin Castellanos (born 1929)*

Agustin Castellanos was born and educated in Cuba and was already a distinguished cardiologist when he immigrated to the USA. He has

---

* Lüderitz, B. (2001) History. *Journal of Interventional Cardiac Electrophysiology*, **5**, 223 (with permission).

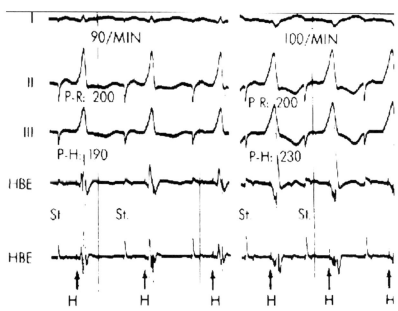

Recording of His bundle electrograms during atrial pacing at increasing rates. Anomalous atrioventricular (AV) conduction appeared at a rate of 90/minute as the atrial rate was increased from 90 to 100 minute. The PR interval remained the same, but the PH interval lengthened in the right panel, the H deflections that appeared after the onset of depolarization in the standard leads. This indicates that part of the ventricles were preexcited ahead of the orthograde His bundle deflections.

His Bundle Electrograms in Two Cases of Wolff-Parkinson-White (Pre-excitation) Syndrome

By Agustin Castellanos, Jr., M.D., Eduardo Chapunoff, M.D., Cesar Castillo, M.D., Orlando Maytin, M.D., and Louis Lemberg, M.D.

SUMMARY

The catheter technic for recording the electrical activity of the specialized conducting system in the human heart showed in two patients studied that ventricular pre-excitation was apparently due to a bypass of the His bundle. Intermediate forms of WPW complexes appeared to be combination beats resulting from the activation of the ventricles through impulses traversing both the His bundle and accessory communications. Preferential iatrogenic activation of an intra-atrial (and perhaps even of an atrioventricular) tract appeared to occur in one of the patients. The patients with the WPW (pre-excitation) syndrome and long histories of paroxysmal arrhythmias were successfully treated with a combination of oral propranolol and implanted (transvenous) demand pacemaker.

made major contributions to the study and comprehension of atrioventricular (AV) conduction and cardiac arrhythmias, based especially on embryology and phylogeny. He was and remains a master electrocardiographer. In the early years of cardiac pacing, atrial synchronous (VAT) in 1963 and demand (VVI) and AV sequential (DVI) later, he was instrumental in analyzing the electrocardiography of pacemaker function. At the turn of the millennium he remains active and is especially involved in studying the effect of pacing on neurocardiogenic syncope.

References
Castellanos, A., Jr., Chapunoff, E., Castillo, C., Maytin, O. & Lemberg, L. (1970) His bundle electrograms in two cases of Wolff–Parkinson–White (preexcitation) syndrome. *Circulation*, **41**, 399–411.
Lüderitz, B. (1984) *Therapie der Herzrhythmusstoerungen*. Springer, Berlin.

Demetrio Sodi-Pallares (1913–2003)*

Dr. Demetrio Sodi-Pallares was born 1913 in Mexico City. He was the scion of a very distinguished family of law professors in the University of Mexico. He studied at the National University of Mexico School of Medicine (1929). After graduation, he worked in several Clinical

---

* Lüderitz, B. (2004) History. *Journal of Interventional Cardiac Electrophysiology*, **10**, 93–4 (with permission).

FEBRUARY 1962

THE AMERICAN JOURNAL OF CARDIOLOGY

Clinical Studies

Effects of an Intravenous Infusion of a Potassium-Glucose-Insulin Solution on the Electrocardiographic Signs of Myocardial Infarction

A Preliminary Clinical Report*

DEMETRIO SODI-PALLARES, M.D., F.A.C.C., MARIO R. TESTELLI, M.D.,† BERNARDO L. FISHLEDER, M.D.,
ABDO BISTENI, M.D., GUSTAVO A. MEDRANO, M.D., CHARLOTTE FRIEDLAND, M.D. and
ALFREDO DE MICHELI, M.D.

Mexico City, Mexico

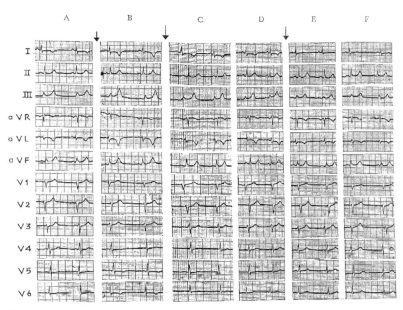

(A) December 26, 1960. (B) December 27, after recurrent pain of 4 hours' duration. (C) Same day after treatment. (D) January 1, 1961. (E) January 8, after treatment. (F) January 11.

Services at the Mexico City General Hospital, such as Endocrinology, Dermatology, and Cardiology, with the guidance of the founder of Mexican cardiology Doctor Ignacio Chavez, at that time Chief of the Cardiology Service in the same hospital and Professor of Cardiology at the University of Mexico. He went to work as a fellow in physiology with Wigger and Wilson in Ann Arbor Michigan. After his return to Mexico, he was able to make a brilliant synthesis of the most outstanding schools of electrocardiography. He learned about the French school from Chavez, the Austrian school of Wenckebach from Brumlinck, and the Anglo-Saxon school of Sir Thomas Lewis from Wilson. The synthesis was the basis of his studies of the electrical process of the activation and its correlation with anatomical, clinical, and physiological findings in patients. In the early 1960s, he was the pioneer in the knowledge of metabolic disorders and the heart, developing the novel treatment with glucose, insulin, and potassium (GKI), a therapy that may play an important role in reducing hospital mortality after acute myocardial infarction (Castellanos, 2003; G.A.J. Gonzalez-Hermosillo, personal communication, 2003).

In 1963, Dr. Demetrio Sodi-Pallares became Vice President and later on President of the National Academy of Medicine (Mexico). He was awarded in 1964 with the Gold Medal of the National Institute of Cardiology. In 1983, he received the Honorary Doctor Degree of Cordoba (Spain), and in 1993 the same Honorary Degree from the University of Alcalá de Henares, Madrid, Spain. Sodi-Pallares published more than 20 books and received more than 100 scientific awards. In particular, his 1945 book *Nuevas Bases de Electrocardiografia* 1st edition, Instituto Nacional (New Basis of Electrocardiography) became very famous. Dr. Sodi-Pallares died in August 12, 2003 at the age of 90 years in Mexico.

A potassium–glucose–insulin treatment had been employed by Sodi-Pallares in clinical cases of myocardial infarction. In most of the patients, the treatment was administered by intravenous infusion of a solution of 40 mEq of KCl and 20 units of regular insulin in 1000 cc of 5 or 10% glucose in water. The speed of infusion was 40 to 60 drops per minute. The major contraindications to the treatment are shock and impairment of renal function. The best results were obtained with the electrocardiographic signs of "acute" and "chronic" injury and ischemia (Bolte & Lüderitz, 1968).

Dr. Sodi-Pallares' investigation was based on the following assumptions:

(a) The normal diastolic polarization of human cardiac fibers equated to a normal K_i^+/K_0^+ gradient, and the decreased diastolic polarization of myocardial infarction to a decreased K_i^+/K_0^+ gradient because of a decreased K_i^+.

(b) The potassium–glucose–insulin treatment by "polarizing" the fiber membranes was expected to modify the electrocardiographic signs of acute myocardial infarction (Sodi-Pallares *et al.*, 1962). The effects of the treatment on arrhythmias and other conditions were obvious. Different arrhythmias disappeared. In regard to ventricular premature beats, their disappearance could simply be due to the diminution or disappearance of the injury zone in which premature beats are known to originate. Dr. Sodi-Pallares also observed favorable results with his treatment in cases of coronary heart disease associated with complete atrioventricular (AV) block, atrial fibrillation, paroxysmal tachycardia, and extrasystoles which had proved "refractory" to the standard treatment (Sodi-Pallares *et al.*, 1960; Sodi-Pallares *et al.*, 1962; Bolte & Lüderitz, 1968).

References
Bolte, H.D. & Lüderitz, B. (1968) Influence of insulin on membrane potentials in potassium deficiency. *Plügers Archiv*, **301**, 254–8.
Castellanos, A. (2003) Demetrio Sodi-Pallares: The man and his thought. In: *Profiles in Cardiology*. (eds J.W. Hurst, C.R. Conti & W.B. Fye), pp. 383–4. Foundation for Advances in Medicine and Science, Mahwah, NJ.
Sodi-Pallares, D., Fishleder, B.L., Cisneros, F. *et al.* (1960) A low sodium, high water, high potassium regimen in the successful management of some cardiovascular diseases. *Canadian Medical Association Journal*, **83**, 243–57.
Sodi-Pallares, D., Testelli, M.R., Fishleder, B.L. *et al.* (1962) Effects of an intravenous infusion of a potassium–glucose–insulin solution on the electrocardiographic signs of myocardial infarction. *The American Journal of Cardiology*, **9**, 166–81.

Melvin M. Scheinman; Catheter ablation for atrial fibrillation and atrial flutter: From DC shocks to radiofrequency current*

Mitsui *et al.* (1978) reported in 1978 on successful transvenous electrocautery of the atrioventricular (AV) connection in mongrel dogs guided by the His electrogram. In 1981, Vedel *et al.* (1979) reported the first AV block induced by an electrode catheter. A patient was undergoing an electrophysiological study following defibrillation, when the defibrillating electrode accidentally came into contact with an electrode catheter in the bundle of His. Gonzales *et al.* (1981) studied this therapeutic electrophysiological procedure in humans. This accidental experiment of Vedel's was refined in 1982 by Gallagher *et al.* (1982), as well as

* Lüderitz, B. (2000) History. *Journal of Interventional Cardiac Electrophysiology*, **4**, 441 (with permission).

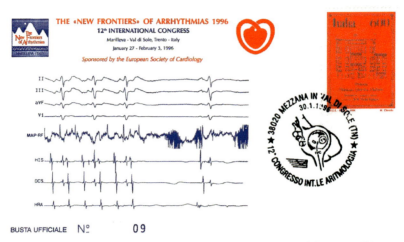

BUSTA UFFICIALE   N.º   09

This first-day issue (*Busta ufficiale*) of the Twelfth International Congress "New Frontiers of Arrhythmia" in Marilleva in 1996, shows how accepted the notion of radiofrequency catheter ablation has become. The issue remembers the great progress made by transcatheter radiofrequency ablation to terminate common atrial flutter, type I, through an approach on the isthmus located between the inferior vena cava, coronary sinus, and tricuspid valve of the reentrant circuit. The artifact present in the MAP-RF (ablation catheter) is due to the radiofrequency. The special postal convalidation synthesizes the type of approach of the RF electrocatheter and the type of atrial flutter circuit, and the typical counterclockwise and subsequent clockwise activation after effective ablation is a marker of a successful ablation. The position of the different catheters can be recognized from the stamp (F. Furlanello, personal communication, 1996).

Scheinman *et al*. (1982) in patients with drug-refractory supraventricular tachycardia. In these patients, a non-invasive percutaneous excision and coagulation of the bundle of His was performed using DC shock via a catheter. This method employs an initial electrophysiological study for diagnostic purposes, and then uses an electrode catheter inserted into the bundle of His. An electrical shock is delivered from an external defibrillator. Generally, this leads to coagulation necrosis in the area around the bundle of His. Throughout this procedure, an external pacemaker provides ventricular pacing support.

References

Gallagher, J.J., Svenson, R.H., Kasell, J.H. *et al*. (1982) Catheter technique for closed-chest ablation of the atrio-ventricular conduction system: A therapeutic alternative for the treatment of refractory supraventricular tachycardia. *New England Journal of Medicine*, **306**, 194–200.

Gonzales, R., Scheinman, M., Margaretten, W. & Rubinstein, M. (1981) Closed-chest electrode-catheter technique for His bundle ablation in dogs. *American Journal of Physiology*, **241**, H283–7.

Mitsui, T., Jima, H., Okamura, K. & Hori, M. (1978) Transvenous electrocautery of the atrioventricular connection guided by the His electrogram. *Japanese Circulation Journal*, **42**, 313–18.

Scheinman, M.M., Morady, F., Hess, D.S. & Gonzales, R. (1982) Transvenous catheter technique for induction of damage to the atrioventricular junction in man. *The American Journal of Cardiology*, **49**, 1013.

Vedel, J., Frank, R., Fontaine, G. & Grosgogeat, Y. (1979) Bloc auriculoventri-culaire intra-Hisien définitif induit au cours d'une exploration endocavitaire droite. *Archives des maladies du coeur et des vaisseaux*, **72**, 107.

John J. Gallagher; Cryosurgical ablation of accessory atrioventricular connections*

A method for correction of the preexcitation syndrome
Gallagher, J.J., Sealy, W.C., Anderson, R.W., Kasell, J., Millar, R., Campbell, R.W.R., Harrison, L., Pritchett, E.L.C., Andrew, A.G. (1977) Cryosurgical ablation of accessory atrioventricular connections. A method for correction of preexcitation syndrome. *Circulation* 55, 471–9.

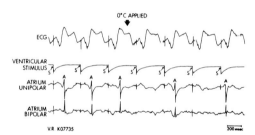

Induction of ventriculo-atrial block with cooling of the accessory pathway. The ventricle is being paced at a cycle length of 400 ms. 1 : 1 VA conduction is present until cooling (0°C) is applied to the accessory pathway. Ventriculo-atrial dissociation follows. Electrocardiogram (ECG) = inverted lead 1; S = stimulus; A = reference atrial electrogram.

---

* Lüderitz, B. (1999) History. *Journal of Interventional Cardiac Electrophysiology*, **3**, 87 (with permission).

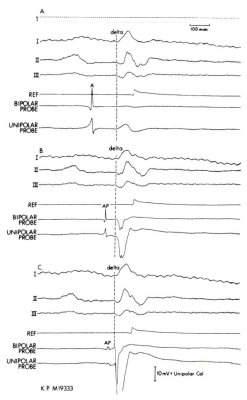

Epicardial recordings from the area of preexcitation. Tracings in panels A–C are, from the top: Electrocardiogram (ECG) leads I–III, a ventricular reference, and bipolar and unipolar recordings from the mapping probe. In panel A, the probe is placed over the atrium continuous to the site of ventricular preexcitation, and records an atrial electrogram (A). In panel B, the probe is on the ventricle and records an electrogram thought to arise in the accessory pathway (AP). In panel C, the probe has been moved 1 cm distal to the AV ring. The previously noted electrogram (AP) merges with the ventricular electrogram. A 10 mV calibration signal is shown to standardize all the unipolar recordings.

Nikos Protonotarios; Naxos disease*

The island of Naxos is situated to the northeast of a group of islands called the Cyclades; Naxos is the largest of these islands. Its population is around 22 000. According to the legend, the Greek island Naxos was the Isle of Dionysos, the god of vines and wine who was born from a thigh

---

* Lüderitz, B. (2003) History. *Journal of Interventional Cardiac Electrophysiology*, **9**, 405–6 (with permission).

Br Heart J 1986;56:321–6

Cardiac abnormalities in familial palmoplantar keratosis

N PROTONOTARIOS, A TSATSOPOULOU, P PATSOURAKOS,
D ALEXOPOULOS, P GEZERLIS, S SIMITSIS, G SCAMPARDONIS
From the Department of Cardiology, 401 Army General Hospital, Athens, Greece

SUMMARY Cardiac abnormalities were identified in patients with familial palmoplantar keratosis. All of them were descended from families on the Greek island of Naxos. Four families were studied and nine cases of palmoplantar keratosis were identified; seven of them showed symptoms and signs of heart disease. Cardiomegaly on chest x ray and electrocardiographic abnormalities were common findings. Three cases had episodes of ventricular tachycardia and a fourth patient died suddenly. All patients with cardiac signs and symptoms showed echocardiographic enlargement of the right ventricle and a right ventricular band; in three the left ventricle was also affected.

Dionysos sits resting on a rock, his leg on the thigh of Ariadne, a pose symbolic of "sacred marriage." (Bronze krater from Derveni, 320–300 BC.)

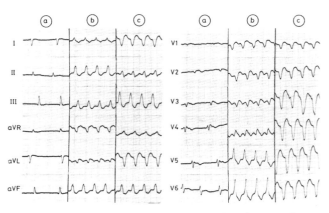

Normal electrocardiogram (ECG) (a) and two episodes of ventricular tachycardia (b and c) in patient 1 (single channel recorder). In (a) there was a QRS axis of > 135 ms, QRS prolongation, low voltage, and T-wave inversion in the precordial leads. In both (b) and (c) the ventricular tachycardia rate is 160 beats per minute and the QRS axes were > 75 ms and > 135 ms, respectively. From Protonotarios, N. *et al.* (1986) (See references)

of his father, Jupiter. The wife of Dionysos was Ariadne, the daughter of Minos. Dr. Nikos Protonotarios and colleagues examined the population of the Greek island of Naxos and described the so-called Naxos disease in 1986 (Protonotarios *et al.*, 1986): A syndrome of arrhythmogenic right ventricular cardiomyopathy (ARVC), non-epidermolytic palmoplantar keratoderma, and woolly hair. The autosomal recessive ARVC of Naxos disease is similar to autosomal dominant ARVC with respect to age and mode of clinical presentation, distribution of right ventricular and left ventricular involvement, electrocardiographic features, natural history, and morphological and histological features. Initial clinical presentation is often with ventricular arrhythmia of left-bundle-branch-block morphology, which suggests a right ventricular origin. All patients originally examined by Nikos Protonotarios and his pediatrician wife Adalena Tsatsopoulou with cardiac abnormalities and familial palmoplantar keratosis were descended from families on the Greek island of Naxos. Cardiomegaly on chest X-ray and electrocardiographic abnormalities were common findings. Three cases had episodes of ventricular tachycardia and a fourth patient died suddenly. All patients with cardiac signs and symptoms showed echocardiographic enlargement of the right ventricle and a right ventricular band; in three the left ventricle was also affected.

The term "Naxos disease" was introduced by G. Fontaine, N. Protonotarios, A. Tsatsopoulou and colleagues with an abstract sent to the

American Heart Association in 1994 (Fontaine *et al.*, 1994). Based on electrocardiography and pathology, the authors suggested that Naxos disease and arrhythmogenic right ventricular dysplasia (ARVD) are two expressions of the same cardiac disorder. It was supposed that in Naxos disease ARVD is genetically transmitted and morbidity associated with palmoplantar keratosis (Fontaine *et al.*, 1994).

Coonar *et al.* (1998) have confirmed autosomal recessive inheritance for Naxos disease and mapped the disorder to chromosome 17q21. Recently, McKoy *et al.* (2000) have shown a two-base pair homozygous deletion in the *plakoglobin* gene in individuals with Naxos disease.

References

Coonar, A.S., Protonotarios, N., Tsatsopoulou, A *et al.* (1998) Gene for arrhythmogenic right ventricular cardiomyopathy with diffuse non-epidermolytic palmoplantar keratoderma and woolly hair (Naxos disease) maps to 17q21. *Circulation*, **97**, 2049–58.

Fontaine, G., Protonotarios, N., Tsatsopoulou, A. *et al.* (1994) Comparisons between Naxos disease and arrhythmogenic right ventricular dysplasia by electrocardiography and biopsy (Abstract). *Circulation*, **90**(2), 3233.

McKoy, G., Protonotarios, N., Crosby, A. *et al.* (2000) Identification of a deletion in *plakoglobin* in arrhythmogenic right ventricular cardiomyopathy with palmoplantar keratoderma and woolly hair (Naxos disease). *Lancet*, **355**, 2119–24.

Protonotarios, N., Tsatsopoulou, A., Patsourakos, P. *et al.* (1986) Cardiac abnormalities in familial palmoplantar keratosis. *British Heart Journal*, **56**, 321–6.

Bernard Lown (born 1921)*

Bernard Lown was born in Lithuania on June 7, 1921, and immigrated with his family to the USA in 1935. He studied medicine at the University of Maine and later at the Johns Hopkins University School of Medicine in Baltimore, where he earned his doctorate in 1945. After training at various hospitals, he held a research position in cardiology at the Peter Bent Brigham Hospital in Boston from 1950 to 1953. After his military service, he continued his internship at the Peter Bent Brigham Hospital and Harvard Medical School in Boston. Lown was then named Director of the Samuel A. Levine Cardiovascular Research Laboratory from 1956 to 1980, and was on staff at the Peter Bent Brigham Hospital. In addition, he also served as Assistant Professor of Medicine at the Harvard School of Public Health from 1961 to 1967. Before he undertook the coordination of a joint study on sudden cardiac death between the USA and the former

---

* Lüderitz, B. (2004) History. *Journal of Interventional Cardiac Electrophysiology*, **10**, 293–4 (with permission).

Bernard Lown (right) and Ewgeni Chazow (Minister of Health of the former USSR) receiving the Nobel Peace Prize in 1985.

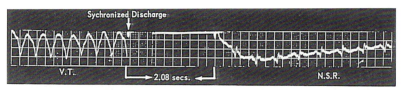

Lown et al., JAMA, Nov. 3, 1962: 548-555

Cardioversion in a patient with ventricular tachycardia (VT). After the defibrillatory discharge of 100 watts, there was an asystolic pause of 2.08 seconds followed by normal sinus rhythm (NSR).

USSR from 1973 to 1981, he was Associate Professor of Cardiology at the Harvard School of Public Health. Bernard Lown is Professor Emeritus of Cardiology at the Harvard School of Public Health and senior physician at Brigham and Women's Hospital in Boston. Cofounder of International Physicians for the Prevention of Nuclear War (IPPNW), he accepted the Nobel Peace Prize on behalf of the organization in 1985.

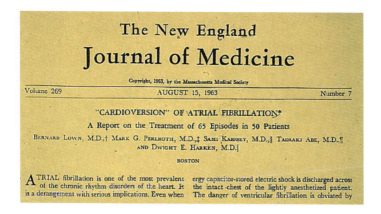

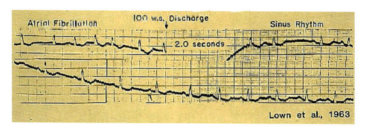

Lown et al., 1963

He developed the DC defibrillator and the cardioverter, and introduced lidocaine as an antiarrhythmic drug. Lown and electrical engineer
Dr. Bernard Berkowitz studied the efficacy and safety of several DC
waveforms in animals, showing that one was consistently effective in
reversing the most intractable episodes of ventricular fibrillation that did
not respond to alternating current. They learned that ventricular fibrillation could be prevented by synchronizing the shock to avoid the vulnerable period of the cardiac cycle, thereby providing a safe method for
reverting tachycardias, a method that Lown called "cardioversion." This
is also true for atrial fibrillation. His recent work demonstrates the role of
psychological and behavioral factors in regulating the heart.

Dr. Lown is the recipient of the George F. Kennan Award, as well
as numerous honorary degrees and other awards both in the USA and
abroad. In 1998, he received the Pioneer in Cardiac Pacing and Electrophysiology Award of the North American Society of Pacing and
Electrophysiology—today the Heart Rhythm Society.

References

Lown, B., Amarasingham, R. & Neumann, J. (1962) New method for terminating
cardiac arrhythmias. Use of synchronized capacitor discharge. *JAMA: The Journal of the American Medical Association*, **182**, 548–55.

Lown, B., Perlroth, M.G., Kaidbey, S., Abe, T. & Harken, D.E. (1963) "Cardioversion" of atrial fibrillation. A report on the treatment of 65 episodes in 50 patients. *New England Journal of Medicine*, **269**, 325–31.

Lüderitz, B. (2002) *History of the Disorders of Cardiac Rhythm*, 3rd edn. Futura, Armonk, NY.

Alan John Camm (born 1947)*

A. John Camm is an Englishman born in Lincolnshire in the north of England. At the 18 years of age he entered the University of London, where he obtained his baccalaureate degree with a physiology major. He went on to study medicine at Guy's Hospital Medical School, receiving his MD in 1971. Afterwards, he worked at Guy's Hospital for 3 years under the tutelage of cardiologist Edgar Sowton, one of the greats of the early days of pacing. Dr. Camm's doctoral thesis was entitled "The application of pacemakers to tachycardia termination," which was accepted by the University of London in 1981.

Professor Camm is a past Chairman of the European Working Group on Arrhythmias, a past President of the British Pacing and Electrophysiology Group, and currently the President of the British Cardiac Society. He is widely recognized for his research and teaching roles at national and international levels. In 2001, the North American Society of Pacing and Electrophysiology awarded him the Distinguished Teacher Award.

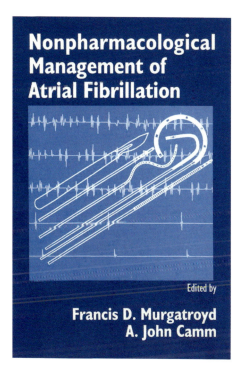

**Nonpharmacological
Management of
Atrial Fibrillation**

Edited by

**Francis D. Murgatroyd
A. John Camm**

Date of Birth: 11 January, 1947, Lincolnshire
Country of Birth: England, United Kingdom
Nationality: very British
Marital Status: married – Joy-Maria
                          (nee Frappell)
Children: Christian John Fielder,
                          Kathryn Lucy Ellis

Queen's Honorary Physician (QHP)
Bachelor of Science (BSc)
Medical Doctor (MD)
Fellow of the Royal College of Physicians (FRCP)
Fellow of the American College of Cardiology (FACC)
Founding Fellow of the European Society of Cardiology
(FESC)
Fellow of the Council of Geriatric Cardiology (FCGC)
Commander of the Venerable Order of St. John of
Jerusalem (CSt.J)
Professor of Clinical Cardiology (British Heart
Foundation), The Prudential Chair of Cardiology,
St. George's Hospital Medical School, University of
London
Honorary Consultant in Cardiology, St. George's
Hospital

Alan John Camm

Camm, together with Dr. S. Saksena, is coediting a major textbook entitled *Electrophysiological Disorders of the Heart* (not yet published).

Reference
Lüderitz, B. (2003) Alan John Camm. In: Profiles in cardiology (eds J.W. Hurst & W.B. Fye). *Clinical Cardiology*, **26**, 251–3.

Heinz Sterz; Transesophageal pacing*

Transesophageal procedures for electrophysiological stimulation were firstly published by the Second Department of Medicine of the Landeskrankenhaus Klagenfurt—Austria (Sterz *et al.*, 1978). The publication by Sterz *et al.* (1978) is the first report describing the successful termination of paroxysmal supraventricular tachycardia in nine patients by rapidly stimulating the left atrium by an esophageal stimulation probe adjacent to the left atrium. This report is also mentioned by Gallagher *et al.* (1982) as the first description of the method.

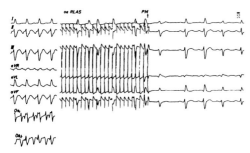

Transesophageal rapid stimulation of the left atrium in atrial tachycardias. Electrocardiograms (ECGs) of a patient before, during, and after transesophageal rapid stimulation of the left atrium (oeRLAS) via an esophageal stimulation catheter. (*Left*) Leads I, II, III, aVR, aVL, and aVF, and two esophageal leads: atrial tachycardia with 2 : 1 atrioventricular (AV) block. (*Center and right* [in a continuous recording]) Rapid left atrial stimulation via the esophageal catheter; artificial induction of atrial fibrillation. After stopping of esophageal rapid left atrial stimulation (oeRLAS): sinus rhythm after the second beat.

References
Gallagher, J.J., Smith, W.M., Kerr, C.R. *et al.* (1982) Esophageal pacing: A diagnostic and therapeutic tool. *Circulation*, **65**, 336–41.

---

* Lüderitz, B. (1998) History. *Journal of Interventional Cardiac Electrophysiology*, **2**, 309 (with permission).

Sterz, H., Prager, H. & Koller, H. (1978) Transösophageale rasche Stimulation des linken Vorhofes zur Elektrotherapie ektoper, tachykarder Vorhofrhythmusstörungen. *Zeilschrift fur Kardiologie*, **87**, 135–8.

Sterz, H., Prager, H. & Koller, H. (1981) Transösophageale Prüfung der Sinusknotenfunktion. *Herzschrittmacher*, **1**, 45–7.

Obituary: Philippe Coumel (1935–2004): A giant of modern clinical electrophysiology* by Samuel Lévy

On March 18, 2004, our colleague and friend Professor Philippe Coumel of Hôpital Lariboisière of Paris, died at age 68 after a terrible disease which had handicapped him for the last few years. French cardiology is deeply saddened as well as the field of cardiac electrophysiology worldwide. His European colleagues and colleagues from all over the world have manifested their sympathy as Philippe was one of the most appreciated and respected opinion leaders in our field. Our first thoughts and condolences are directed towards his family, his wife Claude and his three children, who represent what he cherished most. Aside from his family, Philippe Coumel's life was entirely devoted to cardiology, particularly to the understanding of cardiac arrhythmias. When his interest for this field started in the mid 1960s, as a pupil of Professor Bouvrain, we didn't have any significant insights into arrhythmia mechanisms,

* Lüderitz, B. (2004) History. *Journal of Interventional Cardiac Electrophysiology*, **11**, 77–8 (with permission).

aside from the speculative interpretation of electrocardigrams (ECGs) using ladder diagrams and deductive reasoning. Coumel teamed with Prof. Robert Slama and they have worked together since then, advancing month after month as if explorers of a new continent. They represented a source of admiration for fellows in cardiology all over France, and inspired interest for arrhythmology in the new generation.

While I was a fellow at the University of Miami, my mentor Dr. Agustin Castellanos, familiar with Philippe Coumel's work, most of it written in French at that time, advised me to meet with him during a trip to Paris and to invite him back as a guest of the University of Miami, which I did. A few months later Philippe visited us in Miami and I was very proud to introduce him to Drs. Onkar Narula, Robert Myerburg, and Agustin Castellanos. Back in France, I visited with him regularly at Hôpital Lariboisière, where he was conducting active research using programmed electrical stimulation in patients with supraventricular tachycardias at a time when the focus was mainly on conduction disturbances using His bundle recordings.

Philippe Coumel's contribution to our field is enormous. He was among the first, along with Prof. H.J.J. Wellens, to use programmed electrical stimulation in order to understand the mechanism of paroxysmal supraventricular tachycardias involving the atrioventricular (AV) node, and to describe the reciprocating mechanism of atrioventricular nodal reentrant tachycardias (AVNRT). He described the permanent junctional reciprocating tachycardias (PJRT, referred to in France as *tachycardies de Coumel*), whose mechanism involving a slow conducting concealed accessory connection was subsequently demonstrated by John Gallagher, another major contributor to our field. Philippe Coumel also described catecholamine-induced polymorphic ventricular tachycardias in children. His interest in the autonomous nervous system and its analysis using ambulatory 24-hour recording and computer analysis resulted in the definition of other new syndromes such as vagal atrial fibrillation—occurring at night, during sleep, or after heavy meals—or torsades de pointes—induced by short coupled ventricular extrasystoles in patients with normal QT interval. His contribution to our field is remarkable in many ways, and it is not my purpose to describe in detail what all arrhythmologists are familiar with.

Philippe Coumel is simply irreplaceable. He had an exceptional sense of observation and an intuitive reasoning, which elicited the admiration of all those who had the chance to work with him or to approach him. Until the end, he continued to practice medicine and to visit his patients whom he was eager to help. Philippe Coumel was a man of character, a man of wisdom, and a man of strong convictions, although he was always ready to listen to different opinions. Another aspect of Philippe

Coumel's personality is that he had a hidden sensitiveness, which made him a warm human being to those capable of discovering this aspect and a faithful friend. I had the good fortune to be one of his friends and was privileged with Prof. Fauchier to have shared with him and Claude, his wife, a dinner at the EUROPACE meeting in Paris last December, which he accepted despite the handicap of his illness. He will be remembered by all arrhythmologists as a giant and a pioneer of modern electrophysiology. His friends will miss him and he will remain dear to our hearts.

The best tribute to Philippe comes from my daughter Virginie, aged 24, who asked me why I felt sad. I told her that a friend of mine, Philippe Coumel, had passed away and she said, "It is sad since his name has been with me throughout my childhood."

Samuel Lévy, MD, FACC, FESC, University of Marseille, France

Douglas Peter Zipes (born 1939)*

Douglas Peter Zipes was born in White Plains, New York, on February 27, 1939, and grew up in Pleasantville, New York. After attending Dartmouth Medical School in Hanover, New Hampshire, he transferred to Harvard Medical School, where he graduated 1964 *cum laude*. His

* Lüderitz, B. (2004) History. *Journal of Interventional Cardiac Electrophysiology*, **10**, 1–2 (with permission).

The American Journal of CARDIOLOGY 1971; 28:211 -222

The Contribution of Artificial Pacemaking to Understanding the Pathogenesis of Arrhythmias

DOUGLAS P. ZIPES, MD*

Portsmouth, Virginia

From the Cardiopulmonary Laboratory, Department of Medicine, Naval Hospital, Portsmouth, Va.

Right atrial or ventricular pacing was performed on 36 occasions in 26 patients in an attemptto terminatea variety of tachyarrhythmias. Of 16 episodes of atrial flutter, 13 were terminated successfully; in 9 of the 13, sinus rhythm or the patient's pre-flutter rhythm was restored immediately, whereas in 4 patients, intervening atrial fibrillation or unstable atrial flutter occured. Pacing terminated paroxysmae atrioventricular junctional or paroxysmal atrial tachycardia on 3 occasions; in a fourth patient, this tachyarrhythmia terminated during catheter manipulation. Six episodes of pacemaker-induced ventricular tachycardia were abolished by ventricular pacing. In 2 patients, atrial tachycardia was only transiently suppressed, and in 1 of these patients, d-c cardioversion produced a similar effect. Atrial fibrillation, spontaneously converting to atrial flutter, resulted during pacing for atrial tachycardia with block, the latter arrhythmia returned when the atrial flutter was terminated.

internship and residency in internal medicine and his fellowship in cardiology were at Duke University Medical Center, Durham, North Carolina. After his military service (Navy) he joined the cardiovascular faculty of Indiana, University School of Medicine, Indianapolis, Indiana, where he has been ever since. Within 6 years he became a full professor. In 1994, he achieved the rank of Distinguished Professor, and in 1995 he became Director of the Cardiology Division and the Krannert Institute of Cardiology.

Dr. Zipes is a premier electrophysiologist, researcher, teacher, author, and the only Editor of *The Journal of Cardiovascular Electrophysiology* until 2004. Since May 2004, he has been the Editor of *Heart Rhythm*, the official Journal of the North American Society of Pacing and Electrophysiology (NASPE)–Heart Rhythm Society. For various periods, he was Editor and Chief of *Progress in Cardiology, Contemporary Treatments in Cardiovascular Disease*, and *Cardiology in Review*. He holds five patents, including the synchronous intracardiac cardioverter. He introduced the concept of

Douglas P. Zipes (*middle*) on a private tour to Lhasa, Tibet, in front of the Outpatient Clinic, Traditional Tibetian Medicine, Hospital of the Tibet Autonomous Region. His colleagues (*from right to left*): Warren Jackman, Ralph Lazarra, Douglas P. Zipes, Edward Platia, Berndt Lüderitz.

low-power defibrillation for ventricular arrhythmias, and has been noted for his leadership of multicenter, prospective randomized clinical trials such as the AVID (antiarrhythmic versus implantable defibrillator) trial. He was Chair of the American Board of Internal Medicine Committee, which developed the examination in clinical cardiac electrophysiology, and has been President of the North American Society of Pacing and Electrophysiology. Zipes became President of the American College of Cardiology in March 2001 (Orlando, Florida). Furthermore, as an avid opera supporter for decades, he served as President of the Indianapolis Opera's Board of Directors.

Dr. Zipes published approximately 290 articles in peer-reviewed medical journals, 14 books, and more than 400 review articles or chapters in books or editorials. His comprehensive text book *Cardiac Electrophysiology: From Cell to Bedside* (Coeditor: José Jalife, MD W.B. Saunders, Philadelphia, 2004) appeared recently in its fourth edition.

Douglas P. Zipes has been married to Joan Jacobus for 42 years; he has three children and five grandchildren. As a hobby he (and his wife) are ambitiously writing detective novels.

References
Roberts, W.C. (2003) Douglas Peter Zipes, MD: A conversation with the Editor. *The American Journal of Cardiology*, **91**, 831–56.
Lüderitz, B. (2002) *History of the Disorders of Cardiac Rhythm*, 3rd edn. Futura, Armonk, NY.
Miller, J.M. (2004) Douglas P. Zipes. *Clinical Cardiology*, **27**, 179–80.

CLINICAL STRATEGIES IN THE TREATMENT OF ATRIAL FIBRILLATION
Bonn-Königswinter, Germany, September 25-27, 1997

1) K.H. Kuck, 2) G. Steinbeck, 3) Mrs. Kuck, 4) S. Hohnloser, 5) G. Ayers,
6) J. Brachmann, 7) W. Schoels, 8) Mrs. Akhtar, 9) Mrs. Allessie, 10) L. Jordaens,
11) K. Wheeler, 12) J. Griffin, 13) M. Allessie, 14) M. Akhtar, 15) H. Lüderitz,
16) H.J. Biersack, 17) M. Lesh, 18) A. Waldo, 19) K. Steinbach, 20) T. Schumacher,
21) K. Lenz, 22) B. Schumacher, 23) M. Manz, 24) W. Jung, 25) Mrs. Furlanello,
26) K. Seidl, 27) H. Omran, 28) Mrs. Klein, 29) M. Santini, 30) A.J. Camm,
31) Mrs. Manz, 32) F. Furlanello, 33) H. Klein, 34) M. Lussier, 35) H.J. Trappe,
36) D. Andresen, 37) B. Lüderitz, 38) H. Schilling, 39) S. Lüderitz, 40) S. Lévy,
41) S. Saksena, 42) A. Cappucci, 43) J. Jalife, 44) H.F. Pitschner, 45) G. Breithardt,
46) A. Kirkutis

Appendix 1
History Table

1580	Mercuriale, G. (1530–1606): *Ubi pulsus sit rarus semper expectanda est syncope*
1717	Gerbezius, M. (1658–1718): *Constitutio Anni 1717 a.A.D. Marco Gerbezio Labaco 10. Decem. descripta. Miscellanea-Emphemerides Academiae Naturae Curiosorum. Cent. VII, VIII. 1718: in Appendice*
1761	Morgagni, G.B. (1682–1771): *De sedibus et causis morborum per anatomen indagatis*
1791	Galvani, L. (1737–1798): *De viribus electricitatis in motu musculari commentarius*
1800	Bichat, M.F.X. (1771–1802): *Recherches physiologiques sur la vie et la mort*
1804	Aldini, G. (1762–1834): *Essai theorique et experimental sur le galvanisme, avec une serie d'experiences faites en presence des commissaires de L'Institut National de France, et en divers amphitheatres de Londres*
1827/1846	Adams, R. (1791–1875); Stokes, W. (1804–1878): *Cases of diseases of the heart, accompanied with pathological observations: observations on some cases of permanently slow pulse*
1872	Duchenne de Boulogne, G.B.A. (1806–1875): *De I'electrisation localisee et de son application à la pathologie et à la therapeutique par courants induits et par courants galviniques interrompus et continues*
1882	von Ziemssen, H. (1829–1902): *Studies on the motions of the human heart as well as the mechanical and electrical excitability of the heart and phrenic nerve, observed in the case of the exposed heart of Catharina Serafin*
1890	Huchard, H.: *La Maladie de Adams–Stokes*

1932	Hyman, A.S.: *Resuscitation of the stopped heart by intracardial therapy. II. Experimental use of an artificial pacemaker*
1952	Zoll, P.M.: *Resuscitation of heart in ventricular standstill by external electrical stimulation*
1958	Elmquist, R.; Senning, Å.: *An implantable pacemaker for the heart*
1958	Furman, S.; Robensin, G.: *The use of an intracardiac pacemaker in the correction of total heart block*
1961	Bouvrain, Y.; Zacouto, F.: *L'Entrainement électrosystolique du coeur*
1962	Lown, B. *et al.*: *New method for terminating cardiac arrhythmias*
1963	Nathan, D.A. *et al.*: *An implantable synchronous pacemaker for the long-term correction of complete heart block*
1969	Berkovits, B.V. *et al.*: *Bifocal demand pacing*
1969	Scherlag, B.J. *et al.*: *Catheter technique for recording His bundle activity in man*
1972	Wellens, H.J.J. *et al.*: *Electrical stimulation of the heart in patients with ventricular tachycardia*
1975	Zipes, D.P. *et al.*: *Termination of ventricular fibrillation in dogs by depolarizing a critical amount of myocardium*
1978	Josephson, M.E. *et al.*: *Recurrent sustained ventricular tachycardia*
1978	Funke, H.D.: *First dual-chamber pacemaker*
1980	Mirowski, M. *et al.*: *Termination of malignant ventricular arrhythmias with an implanted automatic defibrillation in human beings*
1981	Breithardt, G. *et al.*: *Non-invasive detection of late potentials in man—a new marker for VT*
1982	Gallagher, J.J. *et al.*: *Catheter technique for closed-chest ablation of the atrioventricular conduction system: A therapeutic alternative for the treatment of refractory supraventricular tachycardia*

1982	Scheinman, M.M. *et al.*: *Transvenous catheter technique for induction of damage to the atrioventricular junction in man*
1986	Lüderitz, B.; Gerckens, U.; Manz, M.: *Antitachycardia pacemaker (Tachylog) and automatic implantable defibrillator (AID): combined use in ventricular tachyarrhythmias*
1987	Borggrefe, M. *et al.*: *High frequency alternating current ablation of an accessory pathway in humans*
1988	Saksena, S.; Parsonnet, V.: *Implantation of a cardioverter-defibrillator without thoracotomy using a triple electrode system*
1991	Jackman, W.M. *et al.*: *Catheter ablation of accessory atrioventricular pathways (Wolff–Parkinson–White syndrome) by radiofrequency current*
1991	Kuck, K.H. *et al.*: *Radiofrequency current catheter ablation of accessory atrioventricular pathways*
1994	Daubert, C. *et al.*: *Permanent atrial resynchronisation by synchronous biatrial pacing in the preventive treatment of atrial flutter associated with high degree interatrial block*
1994	Cazeau, S.; Mugica, J. *et al.*: *Four chamber pacing in dilated cardiomyopathy*
1995	Camm, A.J. *et al.*: *Implantable atrial defibrillator*
1996	Saksena, S. *et al.*: *Dual-site right atrial pacing for atrial fibrillation*
1997	Jung, W.; Lüderitz, B.: *Implantation of an arrhythmia management system for ventricular and supraventricular tachyarrhythmias*
1998	Haïssaguerre, M. *et al.*: *Spontaneous initiation of atrial fibrillation by ectopic beats originating in the pulmonary veins*
1998	Huang, D.T. *et al.*: *Hybrid pharmucologic and ablative therapy: A novel and effective approach for the management of atrial fibrillation*

Appendix 2
NASPE (North American Society of Pacing and Electrophysiology)–Heart Rhythm Society

Past Presidents

Michael E. Cain, MD, 2003–2004
Mark H. Schoenfeld, MD, 2002–2003
Eric N. Prystowsky, MD, 2001–2002
David S. Cannom, MD, 2000–2001
Gerald V. Naccarelli, MD, 1999–2000
David L. Hayes, MD, 1998–1999
Sanjeev Saksena, MD, 1997–1998
John D. Fisher, MD, 1996–1997
Ralph Lazzara, MD, 1995–1996
Nora Goldschlager, MD, 1994–1995
David G. Benditt, MD, 1993–1994
Gerald C. Timmis, MD, 1992–1993
James D. Maloney, MD, 1991–1992
Victor Parsonnet, MD, 1990–1991
Douglas P. Zipes, MD, 1989–1990
Melvin M. Scheinman, MD, 1988–1989
Doris J.W. Escher, MD, 1988
Michael Bilitch, MD, 1987
Paul C. Gillette, MD, 1986–1987
Jerry C. Griffin, MD, 1985–1986
Albert L. Waldo, MD, 1984–1985
Robert G. Hauser, MD, 1983–1984
Bernard Goldman, MD, 1982–1983
Seymour Furman, MD, 1981–1982
J. Warren Harthorne, MD, 1979–1981

Distinguished Scientist Award recipients

Michael R. Rosen, MD, 2004
Andrew L. Wit, PhD, 2003
Charles Antzelevitch, PhD, 2002
Peter J. Schwartz, MD, 2001

Robert J. Myerburg, MD, 2000
Ralph Lazzara, MD, 1999
Maurits A. Allessie, MD, PhD, 1998
J. Thomas Bigger Jr., MD, 1998
Madison S. Spach, MD, 1997
Albert L. Waldo, MD, 1997
James L. Cox, MD, 1996
Gerard M. Guiraudon, MD, 1996
Douglas P. Zipes, MD, 1995
Paul F. Cranefield, MD, PhD, 1994
Michiel J. Janse, MD, 1993
Borys Surawicz, MD, 1992
Mauricio B. Rosenbaum, MD, 1991
Agustin Castellanos, MD, 1990
Maurice Lev, MD, 1989
Godon K. Moe, MD, PhD, 1988
Michel Mirowski, MD, 1987
Will C. Sealy, MD, 1987
Charles Kossmann, MD, 1986
John C. Callaghan, MD, 1985
Wilfred G. Bigelow, MD, 1985
Jack Hopps, 1985
Wilson Greatbatch, PE, 1984
Brian F. Hoffman, MD, 1983
Barouh Berkovits, MSEE, 1982

Distinguished Service Award recipients

Nancy L. Stephenson, BSN, 2004
Barbara K. Krause, MA, 2003
Melanie T. Gura, MSN, CSN, 2002
Daniel E. Nickelson, 2001
Wilton W. Webster, BSME, 2000
Leonard A. Cobb, MD, 1999
Carol J. McGlinchey, 1998
Jennifer A. Fraser, RN, 1996
Jacques Mugica, MD, 1995
Susan L. Song, RN, BSN, 1994
J. Warren Harthorne, MD, 1992
Henry D. McIntosh, MD, 1991
Victor Parsonnet, MD, 1990
Seymour Furman, MD, 1989
J. Walter Keller, BSEE, MS, 1988

Jean-J. Welti, MD, 1986
Earl Bakken, 1985
William P. Murphy, MD, 1985
David Link, 1980

Distinguished Teacher Award recipients

Fred Morardy, MD, 2004
George J. Klein, MD, 2003
Eric N. Prystowsky, MD, 2002
A. John Camm, MD, 2001
Hein J.J. Wellens, MD, 2000
John J. Gallagher, MD, 1999
Nora Goldschlager, MD, 1998
Howard B. Burchell, MD, 1997
Mark E. Josephson, MD, 1996
Leonard S. Dreifus, MD, 1995
Charles Fisch, MD, 1994
S. Serge Barold, MD, 1992

Pioneer in Cardiac Pacing and Electrophysiology recipients

Michel Haïssaguerre, MD, 2004
John D. Fisher, MD, 2003
Jeremy N. Ruskin, MD, 2002
Mark E. Josephson, MD, 2001
Warren Jackman, MD, 2000
Kenneth B. Stokes, B Ch, 2000
Masood Akhtar, MD, 1999
Victor Parsonnet, MD, 1999
Paul C. Gillette, MD, 1998
Bernard Lown, MD, 1998
Philippe Coumel, MD, 1997
Melvin M. Scheinman, MD, 1996
Hein J.J. Wellens, MD, 1995
Herman K. Hellerstein, MD, 1993
Anthony N. Damato, MD, 1992
Nicholas P.D. Smyth, MD, 1992
Leonardo Cammilli, MD, 1991
William W.L. Glenn, MD, 1990
Doris J.W. Escher, MD, 1989
Benjamin J. Scherlag, PhD, 1989
Paul M. Zoll, MD, 1989

President's Award recipients

Peter M. Spooner, PhD, 2004
David S. Cannom, MD, 2002
Morton M. Mower, MD, 2000
Glenn Rahmoeller, MS, 1986
Doris J.W. Escher, MD, 1985
J. Warren Harthorne, MD, 1984

Heart Rhythm Society Distinguished Achievement Award recipients

Arthur J. Moss, MD, 2000
Seymour Furman, MD, 1996

Appendix 3
Name Index